Principles and Applications of NanoMEMs Physics

Principles and Applications of NanoMEMs Physics

By

H.R. Dogra

2008

CYBER TECH PUBLICATIONS
4264/3, Ansari Road, Daryaganj, New Delhi-110002 (India)
Ph.: 011-23244078. Fax: 011-23280028
E-Mail: cyberpublicationsdelhi@yahoo.com

PRINCIPLES AND APPLICATIONS OF NANOMEMS PHYSICS

H.R. Dogra

First Edition 2008

Published by :

G.S. Rawat for **Cyber Tech Publications**
4264/3, Ansari Road, Daryaganj, New Delhi-110002 (India)
Ph.: 011-23244078. Fax: 011-23280028
E-Mail: cyberpublicationsdelhi@yahoo.com

Preface

The field of Nanotechnology, which aims at exploiting advances in the fabrication and controlled manipulation of nanoscale objects, is attracting worldwide attention. This attention is predicated upon the fact that obtaining early supremacy in this field of miniaturization may well be the key to dominating the world economy of the 21st century, and beyond. NanoMEMS exploits the convergence between nanotechnology and microelectromechanical systems (MEMS) brought about by advances in the ability to fabricate nanometer-scale electronic and mechanical device structures. In this context, NanoMEMS-based applications will be predicated upon a multitude of physical phenomena, e.g., electrical, optical, mechanical, magnetic, fluidic, quantum effects and mixed domain.

This book presents the first unified exposition of the physical principles at the heart of NanoMEMS-based devices and applications. In particular, after beginning with a comprehensive presentation of the fundamentals and limitations of nanotechnology and MEMS fabrication techniques, the book addresses the physics germane to this dimensional regime, namely, quantum wave-particle phenomena, including, the manifestation of charge discreteness, quantized electrostatic actuation, and quantum wave phenomena, including, quantized electrical conductance, quantum interference, Luttinger liquids, quantum entanglement, superconductivity and cavity quantum electrodynamics. Potential building blocks are also addressed for NanoMEMS applications, including, nanoelectromechanical quantum circuits and systems such as charge detectors, the which-path electron interferometer, and the Casimir oscillator, as well as a number of quantum computing implementation paradigms. Finally, NanoMEMS applications in photonics are addressed, including nanophotonic light sources and plasmonic devices.

It is being written specially for advanced undergraduate/ beginning graduate students in Physics, Electrical Engineering, Materials Science, and Mechanical Engineering, faculty developing courses emphasizing physics and applications of nanotechnology, nanotechnology researchers engaged in analyzing, modeling, and designing NanoMEMS-based devices, circuits and systems. Hope that this book will be informative enough to the readers and suggestions regarding the book are always welcome.

—Author

Contents

1

Introduction to NanoMEMS Physics

A host of novel applications and new physics could be unleashed as microelectromechanical systems shrink towards the nanoscale.

In the late 1950s visionary physicist Richard Feynman issued a public challenge by offering $1000 to the first person to create an electrical motor "smaller than 1/64th of an inch". Much to Feynman's consternation the young man who met this challenge, William McLellan, did so by investing many tedious and painstaking hours building the device by hand using tweezers and a microscope.

McLellan's motor now sits in a display case at the California Institute of Technology and has long since ceased to spin. Meanwhile, in the field that Feynman hoped to incite, the wheels are turning - both figuratively and literally - in many university and corporate laboratories, and even on industrial production lines. Indeed, the field of microelectromechanical systems (MEMS), which became firmly established in the mid-1980s, has now matured to the point where we can be rather blasé about the mass production of diminutive motors that are hundreds of times smaller than McLellan's original. Along the way the MEMS community has developed some truly intriguing products - from digital projectors that contain millions of electrically driven micromirrors to microscale motion sensors that sit in cars ready to deploy airbags.

A whole new realm of interconnected microsensors and instruments is emerging from the minds and laboratories of the scientists and engineers engaged in this research. Their devices are being assigned to a myriad of remote outposts, from the depths of the sea and the Earth's crust, to the far-flung regions of space and distant planets. Moreover, the robustness and low cost of such remote microsensors are helping to provide an avalanche of information about our physical surroundings.

MEMS represent the marriage of semiconductor processing to mechanical engineering - at a very small scale. And it is a field that has grown enormously during the past decade. Numerous companies - from the semiconductor giants to fledgling start-ups - are all now scrambling for a piece of the action at the microscale. Yet very little has been done with MEMS at sizes below one micrometre. This stands in striking contrast to the recent developments in mainstream microelectronics where chips are now mass-produced with features as small as 0.18 microns. Indeed, SEMATECH - a think-tank for a consortia of semiconductor companies in the US - predicts that the minimum feature size will shrink to 70 nm by 2010.

In the face of these achievements, and the advances they are expected to bring to mainstream electronics, the time is ripe for a concerted exploration of nanoelectromechanical systems (NEMS) - i.e. machines, sensors, computers and electronics that are on the nanoscale. Such efforts are under way in my group at Caltech, and in several others around the world. The potential payoffs are likely to be enormous and could benefit a diverse range of fields, from medicine and biotechnology to the foundations of quantum mechanics. In this article I highlight a few of the most exciting promises of NEMS, and the challenges that must be faced to attain them.

What is an Electromechanical System?

One of the earliest reported electromechanical devices was built in 1785 by Charles-Augustin de Coulomb to measure electrical charge. His electrical torsion balance consisted of two spherical metal balls - one of which was fixed, the other attached to a moving rod - that acted as capacitor plates, converting a difference in charge between them to an attractive force. The device illustrates the two principal components common to most electromechanical systems irrespective of scale: a mechanical element and transducers.

The mechanical element either deflects or vibrates in response to an applied force. To measure quasi-static forces, the element typically has a weak spring constant so that a small force can deflect it by a large amount. Time-varying forces are best measured using low-loss mechanical resonators that have a large response to oscillating signals with small amplitudes.

Many different types of mechanical elements can be used to sense static or time-varying forces. These include the torsion balance (used by Coulomb), the cantilever (now ubiquitous in scanning probe microscopy) and the "doubly clamped" beam, which is fixed at both ends. In pursuit of ultrahigh sensitivity, even more intricate devices are used, such as compound resonant structures that possess complicated transverse, torsional or longitudinal modes of vibration. These complicated modes can be used to minimize vibrational losses, in much the same way that the handle of a tuning fork is positioned carefully to reduce losses.

The transducers in MEMS and NEMS convert mechanical energy into electrical or optical signals and vice versa. However, in some cases the input transducer simply keeps the mechanical element vibrating steadily while its characteristics are monitored as the system is perturbed. In this case such perturbations, rather than the input signal itself, are precisely the signals we wish to measure. They might include pressure variations that affect the mechanical damping of the device, the presence of chemical adsorbates that alter the mass of the nanoscale resonator, or temperature changes that can modify its elasticity or internal strain. In these last two cases, the net effect is to change the frequency of vibration.

TOWARDS ROUTINE MANUFACTURE AT THE NANOSCALE

NEMS must overcome a final important hurdle before nanoscale machines, sensors and electronics emerge from industrial production lines. Put simply, when we combine state-of-the-art processes from two disparate fields - nanolithography and MEMS micromachining - we increase the chances that something will go awry during manufacturing. Fortunately, sustained and careful work is beginning to solve these problems and is revealing the way to build robust, reliable NEMS. Given the remarkable success of microelectronics, it seems clear that such current troubles will ultimately become only of historical significance.

But there is a special class of difficulties unique to NEMS that cannot be so easily dismissed. NEMS can respond to masses approaching the level of single atoms or molecules. However, this sensitivity is a double-edged sword. On the one hand it offers major advances in mass spectrometry; but it can also make device reproducibility troublesome, even elusive. For example, at Caltech we have found that it places extremely stringent requirements on the cleanliness and precision of nanofabrication techniques.

SOME APPLICATIONS OF NEMS

Ultimately, NEMS could be used across a broad range of applications. At Caltech we have used NEMS for metrology and fundamental science, detecting charges by mechanical methods and in thermal transport studies on the nanoscale (see Schwab *et al.* in further reading). In addition, a number of NEMS applications are being pursued that might hold immense technological promise.

In my opinion, most prominent among these is magnetic resonance force microscopy (MRFM). Nuclear magnetic resonance was first observed in 1946 by Edward Purcell, Felix Bloch and their collaborators, and is now routinely used for medical imaging. The technique exploits the fact that most nuclei have an intrinsic magnetic moment or "spin" that can interact with an applied magnetic field. However, it takes about 10^{14}-10^{16} nuclei to generate a measurable signal. This limits the resolution that can be attained in state-of-the-art magnetic resonance imaging (MRI) research laboratories to about 10 μm. Meanwhile, the typical resolution achievable in hospitals is about 1 mm.

One would assume then that the detection of individual atoms using MRI is only a distant dream. However, in 1991 John Sidles of the University of Washington at Seattle proposed that mechanical detection methods could lead to nuclear magnetic resonance spectrometry that would be sensitive to the spin of a single proton. Achieving this degree of sensitivity would be a truly revolutionary advance, allowing, for example, individual biomolecules to be imaged with atomic-scale resolution in three dimensions.

Magnetic resonance force microscopy (MRFM) could thus have an enormous impact on many fields, ranging from molecular biology to materials science. The technique was first demonstrated in 1992 by Dan Rugar and co-workers at IBM's Almaden Research Center, and was later confirmed by Chris Hammel at the Los Alamos National Lab in collaboration with my group at Caltech, and others.

Like conventional magnetic resonance, MRFM uses a uniform radio-frequency field to excite the spins into resonance. A nanomagnet provides a magnetic field that varies so strongly in space that the nuclear-resonance condition is satisfied only within a small volume, which is about the size of atom. This magnet also interacts with the resonant nuclear spins to generate a tiny "back action" force that causes the cantilever on which the nanomagnet is mounted to vibrate. For a single resonant nucleus, the size of this force is a few attonewtons (10^{-18} N) at the most. Nonetheless, Thomas Kenny's group at Stanford, in collaboration with Rugar's group at IBM, has demonstrated that such minute forces are measurable.

By scanning the tip over a surface, a 3-D map of the relative positions of resonating atoms can be created. Although Rugar and co-workers detected a signal from some 10^{13} protons in their early experiments, the sensitivity still exceeded that of conventional MRI methods.

In another area of research, Clark Nguyen and co-workers at the University of Michigan are beginning to demonstrate completely mechanical components for processing radio-frequency signals.

With the advent of NEMS, several groups are investigating fast logic gates, switches and even computers that are entirely mechanical. The idea is not new. Charles

Babbage designed the first mechanical computer in the 1820s, which is viewed as the forerunner to the modern computer. His ideas were abandoned in the 1960s when the speed of nanosecond electronic logic gates and integrated circuits vastly outperformed moving elements. But now that NEMS can move on timescales of a nanosecond or less, the established dogma of the digital electronic age needs careful re-examination.

TO THE QUANTUM LIMIT - AND BEYOND

The ultimate limit for nanomechanical devices is operation at, or even beyond, the quantum limit. One of the most intriguing aspects of current nanomechanical devices is that they are already on the verge of this limit. The key to determining whether NEMS are in this domain is the relationship between the thermal energy, $k_B T$, and the quantity hf_0, where k_B is the Boltzmann constant, h is the Planck constant, f_0 is the fundamental frequency of the mechanical resonator and T is its temperature.

When the temperature of the device is low and its frequency is sufficiently high that hf_0 greatly exceeds $k_B T$, then any thermal fluctuations will be smaller than the intrinsic quantum noise that affects the lowest vibrational mode. In this limit, the mean square amplitude of the vibration can be quantized and can only assume values that are integral multiples of $hf_0 Q/2k_{eff}$. A full exploration of this quantum domain must wait for crucial technological advances in ultrasensitive transducers for NEMS that will enable us to measure tiny displacements at microwave frequencies.

In spite of this significant challenge, we should begin to see signs of quantum phenomena in nanomechanical systems in the near future. Even the first NEMS resonators produced back in 1994 operated at sufficiently high frequencies that, if cooled to 100 mK, only about 20 vibrational quanta would be excited in the lowest fundamental mode. Such temperatures are readily reached using a helium dilution refrigerator. So the question that comes to mind is whether quantized amplitude jumps can be observed in a nanoscale resonating device?

If so, one should be able to observe discrete transitions as the system exchanges quanta with the outside world. At this point, the answer to the question seems to be that such jumps should be observable if two important criteria can be met. The first is that the resonator must be in a state with a definite quantum number. In general, transducers measure the position of the resonator, rather than the position squared.

The continual interaction between such a "linear transducer" and the quantum system prevents the resonator from being in a state characterized by a discrete number of quanta. Transducers that measure the position squared were discussed in 1980 by Carlton Caves, now at the University of New Mexico, and co-workers at Caltech in a pioneering paper on quantum measurements with mechanical systems (see Caves in further reading). And it now seems possible to transfer their ideas to NEMS.

The second criterion is more problematic. The transducer must be sensitive enough to resolve a single quantum jump. Again, ultrahigh sensitivity to displacements is the key needed to unlock the door to this quantum domain.

A simple estimate shows that we must detect changes in the mean square displacement as small as 10^{-27} m^2 to observe such quantum phenomena. Is it possible to achieve this level of sensitivity? My group at Caltech has recently made significant progress towards new ultra-sensitive transducers for high-frequency NEMS - and we are currently only a factor of 100 or so away from such sensitivity.

In related work, last year Keith Schwab, Eric Henriksen, John Worlock and I investigated the quantum limit, where $hf_0 >> kT$, for the first time in thermal-transport experiments using nanoscale beams fabricated from silicon nitride (figure 6). As the smallest features on the devices are scaled down in size, the energy spacing between the phonons - the quanta of vibrational energy - increases. When the temperature is lowered, fewer and fewer of these modes of vibration (or phonons) remain energetically accessible. Effectively, this means that most of them cannot participate in thermal transport. Indeed, in a beam that is small enough, only four phonon modes can transport energy between the system and its surroundings.

We found that the thermal conductance in this regime becomes quantized. In other words, each phonon mode that transports energy can only provide a maximum thermal conductance given by $\frac{1}{4}k^2T/6h$. Quantum mechanics thus places an upper limit on the rate at which energy can be dissipated in small devices by vibrations.

In spite of the complications encountered at the quantum level, the rewards in terms of intriguing physics will be truly significant. Force and displacement measurements at this limit will open new horizons in science at the molecular level, new devices for quantum computation, and the possibility of being able to control the thermal transport by individual phonons between nanomechanical systems or between a system and its environment.

Once we have passed into this realm of quantum mechanics, the division between quantum optics and solid-state physics becomes increasingly blurred. Many of the same physical principles governing the manipulation of light at the level of individual photons will come into play for both the mechanical and thermal properties of nanoscale systems.

FUTURE OUTLOOK

NEMS offer unprecedented and intriguing opportunities for sensing and fundamental measurements. Both novel applications and fascinating physics will undoubtedly emerge from this new field, including single-spin magnetic resonance and phonon counting using mechanical devices.

To take full advantage of these systems we will have to stretch our imaginations, as well as our current methods and "mindsets" in micro- and nanoscale science and technology.

But there remains a gap between today's NEMS devices that are sculpted from bulk materials and those that will ultimately be built atom by atom.

In the future, complex molecular-scale mechanical devices will be mass-produced by placing millions of atoms with exquisite precision or by some form of controlled self-assembly. This will be true nanotechnology. Nature has already mastered such remarkable feats of atomic assembly, forming molecular motors and machinery that can transport biochemicals within cells or move entire cells.

Clearly, for us to attain such levels of control and replication will take sustained effort, involving a host of laboratories. Meanwhile, in the shorter term, NEMS are clearly destined to provide much of the crucial scientific and engineering foundation that will underlie future nanotechnology.

NEMS in Action

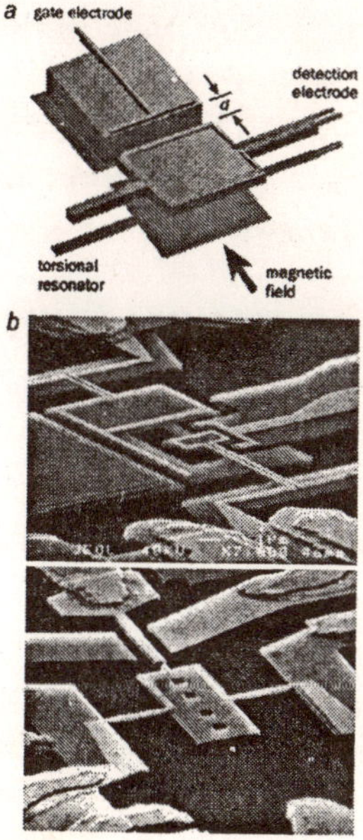

Figure: At the heart of the electrometer is a mechanical resonator made of silicon-on-insulator that rotates when a radio-frequency current is passed through the loop-shaped gold electrode in the presence of a strong magnetic field. The resulting torsional motion generates an electric field that is measured by the detection electrode. When a charge is placed on the gate electrode, the resulting electric field alters the frequency with which the resonator rotates. (b) First- and second-generation mechanical electrometers built at Caltech.

In general, the output of an electromechanical device is the movement of the mechanical element. There are two main types of response: the element can simply deflect under the applied force or its amplitude of oscillation can change (figure 3). Detecting either type of response requires an output or readout transducer, which is often distinct from the input one. In Coulomb's case, the readout transducer was "optical" - he simply used his eyes to record a deflection. Today mechanical devices contain transducers that are based on a host of physical mechanisms involving piezoelectric and magnetomotive effects, nanomagnets and electron tunnelling, as well as electrostatics and optics.

THE BENEFITS OF NANOMACHINES

Nanomechanical devices promise to revolutionize measurements of extremely small displacements and extremely weak forces, particularly at the molecular scale. Indeed with surface and bulk nanomachining techniques, NEMS can now be built with masses approaching a few attograms (10^{-18} g) and with cross-sections of about 10 nm (figure 4). The small mass and size of NEMS gives them a number of unique attributes that offer immense potential for new applications and fundamental measurements.

Mechanical systems vibrate at a natural angular frequency, w_0, that can be approximated by $w_0 = (k_{eff}/m_{eff})^{1/2}$, where k_{eff} is an effective spring constant and m_{eff} is an effective mass. (Underlying these simplified "effective" terms is a complex set of elasticity equations that govern the mechanical response of these objects.) If we reduce the size of the mechanical device while preserving its overall shape, then the fundamental frequency, w_0, increases as the linear dimension, l, decreases. Underlying this behaviour is the fact that the effective mass is proportional to l^3, while the effective spring constant is proportional to l. This is important because a high response frequency translates directly to a fast response time to applied forces. It also means that a fast response can be achieved without the expense of making stiff structures.

Resonators with fundamental frequencies above 10 GHz (10^{10} Hz) can now be built using surface nanomachining processes involving state-of-the-art nanolithography at the 10 nm scale (see box). Such high-frequency mechanical devices are unprecedented and open up many new and exciting possibilities. Among these are ultralow-power mechanical signal processing at microwave frequencies and new types of fast scanning probe microscopes that could be used in fundamental research or perhaps even as the basis of new forms of mechanical computers.

A second important attribute of NEMS is that they dissipate very little energy, a feature that is characterized by the high quality or Q factor of resonance. As a result, NEMS are extremely sensitive to external damping mechanisms, which is crucial for building many types of sensors. In addition, the thermomechanical noise, which is analogous to Johnson noise in electrical resistors, is inversely proportional to Q. High Q values are therefore an important attribute for both resonant and deflection sensors, suppressing random mechanical fluctuations and thus making these devices highly sensitive to applied forces. Indeed, this sensitivity appears destined to reach the quantum limit.

Typically, high-frequency electrical resonators have Q values less than several hundred, but even the first high-frequency mechanical device built in 1994 by Andrew Cleland at Caltech was 100 times better. Such high quality factors are significant for potential applications in signal processing.

The small effective mass of the vibrating part of the device - or the small moment of inertia for torsional devices - has another important consequence. It gives NEMS an astoundingly high sensitivity to additional masses - clearly a valuable attribute for a wide range of sensing applications. Recent work by Kamil Ekinci at

Caltech supports the prediction that the most sensitive devices we can currently fabricate are measurably affected by small numbers of atoms being adsorbed on the surface of the device.

Meanwhile, the small size of NEMS also implies that they have a highly localized spatial response. Moreover, the geometry of a NEMS device can be tailored so that the vibrating element reacts only to external forces in a specific direction. This flexibility is extremely useful for designing new types of scanning probe microscopes.

NEMS are also intrinsically ultralow-power devices. Their fundamental power scale is defined by the thermal energy divided by the response time, set by Q/w_0. At 300 K, NEMS are only overwhelmed by thermal fluctuations when they are operated at the attowatt (10^{-18} W) level. Thus driving a NEMS device at the picowatt (10^{-12} W) scale provides signal-to-noise ratios of up to 10^6. Even if a million such devices were operated simultaneously in a NEMS signal processor, the total power dissipated by the entire system would still only be about a microwatt. This is three or four orders of magnitude less than the power consumed by conventional electronic processors that operate by shuttling packets of electronic charge rather than relying on mechanical elements.

Another advantage of MEMS and NEMS is that they can be fabricated from silicon, gallium arsenide and indium arsenide - the cornerstones of the electronics industry - or other compatible materials. As a result, any auxiliary electronic components, such as transducers and transistors, can be fabricated on the same chip as the mechanical elements. Patterning NEMS so that all the main internal components are on the same chip means that the circuits can be immensely complex. It also completely circumvents the insurmountable problem of aligning different components at the nanometre scale.

HOW TO MAKE MEMS AND NEMS

Over the past six years, at my lab and elsewhere, new techniques have been developed for patterning freely suspended 3-D semiconductor structures. These techniques apply to bulk silicon, epitaxial silicon and silicon-on-insulator heterostructures, as well as to systems based on gallium arsenide and indium arsenide.

(a) In its simplest form, the procedure begins with a heterostructure that contains structural (red) and sacrificial layers (blue) on a substrate (yellow).

(b) Masks on top of this substrate are patterned by a combination of optical and electron-beam lithography, followed by a thin-film deposition processes. The resulting mask (black) protects the material beneath it during the next stage.

(c) Unprotected material around the mask is then etched away using a plasma process.

(d) Finally, a local chemically selective etch step removes the sacrificial layer from specific regions to create freely suspended nanostructures that are both thermally and mechanically isolated.

In typical devices this entire procedure might be repeated several times and combined with various deposition processes to give complicated mechanical nanostructures. The flexibility of the process allows complex suspended structures with lateral dimensions down to a few tens of nanometres to be fabricated. Moreover, complex transducers can be incorporated for control and measurement purposes. Epitaxial growth means that the thickness of the layers can be controlled with atomic precision. In principle, the fabricated devices can be just a few layers thick.

CHALLENGES FOR NEMS

Processes such as electron-beam lithography and nanomachining now enable semiconductor nanostructures to be fabricated below 10 nm. It would appear that the technology exists to build NEMS. So what is holding up applications? It turns out that there are three principal challenges that must be addressed before the full potential of NEMS can be realized: communicating signals from the nanoscale to the macroscopic world; understanding and controlling mesoscopic mechanics; and developing methods for reproducible and routine nanofabrication.

NEMS are clearly very small devices that can deflect or vibrate within an even smaller range during operation. For example, the deflection of a doubly clamped beam varies linearly with an applied force only if it is displaced by an amount that typically corresponds to a few per cent of its thickness.

For a beam 10 nm in diameter, this translates to displacements that are only a fraction of a nanometre. Building transducers that are sensitive enough to allow information to be transferred accurately at this scale requires reading out positions with a far greater precision. A further difficulty is that the natural frequency of this motion increases with decreasing size. So the ideal NEMS transducer must ultimately be capable of resolving displacements in the 10^{-15}-10^{-12} m range and be able to do so up to frequencies of a few gigahertz. These two requirements are truly daunting, and much more challenging than those faced by the MEMS community so far.

To compound the problem, some of the transducers that are mainstays of the micromechanical realm are not applicable in the nanoworld. Electrostatic transduction, the staple of MEMS, does not scale well into the domain of NEMS. Nanoscale electrodes have capacitances of about 10^{-18} farad and less. As a result, the many other, unavoidable parasitic impedances tend to dominate the "dynamic" capacitance that is altered by the device motion.

Meanwhile optical methods, such as simple beam-deflection schemes or more sophisticated optical and fibre-optic interferometry - both commonly used in scanning probe microscopy to detect the deflection of the probe - generally fail beyond the so-called diffraction limit. In other words, these methods cannot easily be applied to objects with cross-sections much smaller than the wavelength of light.

For fibre-optic interferometry, this breakdown can occur even earlier, when devices are shrunk to a fraction of the diameter of the fibre.

Conventional approaches thus appear to hold little promise for high-efficiency transduction with the smallest of NEMS devices. Nonetheless, there are a host of intriguing new concepts in the pipeline. These include techniques that are based on integrated near-field optics, nanoscale magnets, high-electron-mobility transistors, superconducting quantum interference devices and single-electron transistors - to name just a few.

THE ROLE OF SURFACE PHYSICS

One of the keys to realizing the potential of NEMS is to achieve ultrahigh quality factors. This overarching theme underlies most areas of research, with the possible exception of non-resonant applications. However, both intrinsic and extrinsic properties limit the quality factor in real devices. Defects in the bulk material and interfaces, fabrication-induced surface damage and adsorbates on the surfaces are among the intrinsic features that can dampen the motion of a resonator. Fortunately, many of these effects can be suppressed through a careful choice of materials, processing and device geometry.

Extrinsic effects - such as air resistance, clamping losses at the supports and electrical losses mediated through the transducers - can all be reduced by careful engineering. However, certain loss mechanisms are fundamental and ultimately limit the maximum attainable quality factors. These processes include thermoelastic damping that arises from inelastic losses in the material.

One aspect in particular looms large: as we shrink MEMS towards the domain of NEMS, the device physics becomes increasingly dominated by the surfaces. We would expect that extremely small mechanical devices made from single crystals and ultrahigh-purity heterostructures would contain very few defects, so that energy losses in the bulk are suppressed and high quality factors should be possible.

For example, Robert Pohl's group at Cornell University, and others, has shown that centimetre-scale semiconductor MEMS can have Q factors as high as 100 million at cryogenic temperatures. But my group at Caltech has shown repeatedly over the past seven years that this value decreases significantly - by a factor of between 1000 and 10 000 - as the devices are shrunk to the nanometre scale.

The reasons for this decrease are not clear at present. However, the greatly increased surface-to-volume ratio in NEMS, together with the non-optimized surface properties, is the most likely explanation.

This can be illustrated by considering a NEMS device fabricated using state-of-the-art electron-beam lithography. A silicon beam 100 nm long, 10 nm wide and

10 nm thick contains only about 5 x 10^5 atoms, with some 3 x 10^4 of these atoms residing at the surface. In other words, more than 10% of the constituents are surface or near-surface atoms.

It is clear that these surface atoms play a central role, but understanding exactly how will take considerable effort. My group and others - at IBM's Almaden Research Center, Stanford University, the University of California at Santa Barbara and Cornell University, all in the US, together with Ludwig-Maximilians University in Munich, Germany - are currently exploring this crucial issue.

Ultimately, as devices become ever smaller, macroscopic mechanics will break down and atomistic behaviour will emerge. Indeed, molecular dynamics simulations, such as those performed by Robert Rudd and Jeremy Broughton at the Naval Research Laboratory in Washington DC on idealized structures just a few tens of atoms thick, would appear to support this idea.

NANOELECTROMECHANICAL SYSTEMS

NEMS or nanoelectromechanical systems are similar to MEMS but smaller. They hold promise to improve abilities to measure small displacements and forces at a molecular scale, and are related to nanotechnology.

There are two approaches most researchers accept as standard paths to NEMS. The top-down approach can be summarized as "a set of tools designed to build a smaller set of tools". For example, a millimeter sized factory that builds micrometer sized factories which in turn can build nanometer sized devices.

The other approach is the bottom-up approach, and can be thought of as putting together single atoms or molecules until a desired level of complexity and functionality has been achieved in a device. Such an approach may utilize molecular self-assembly or mimic molecular biology systems.

A combination of these approaches may also be used, in which nanoscale molecules are integrated into a top-down framework. One such example is the carbon nanotube nanomotor.

University of Michigan scientists have received a pilot grant of nearly $1.3 million from the Defense Advanced Research Projects Agency to test whether nanoparticles can solve a pressing problem in battle zones like Iraq: how to administer sustained, safe doses of the most effective painkillers to injured soldiers, long before they can reach expert medical help.

2

Molecular Mechanics and Molecular Dynamics Analysis

Over the past two years at the Materials and Process Simulation Center, we have been developing simulation approaches for studying the molecular nanomachine designs pioneered by Drexler and Merkle. These nanomachine designs, such as planetary gears and neon pump are described with atomistic details and involves up to 10000 atoms.

With the Dreiding and Universal Force Fields, we have optimized the structures of the two Planetary Gear design and Neon Pump. At the Fourth Foresight conference, we reported rotational impulse dynamics studies of the first and second generation designs of Planetary Gears undergoing very high frequency rotational motions. We will explore stability of these designs in the lower frequency regimes which require long time simulations. We will report the molecular mechanics and molecular dynamics simulations performed on these model systems. We explore the following modes in these studies:

1. Impulse mode.
2. Constant angular velocity -Perpetual rotation.
3. Constant torque -Acceleration from rest.

INTRODUCTION

In order to lay a foundation for creating a technology for manufacturing at the nanoscale level, Eric Drexler, Ralph Merkle, and collaborators[1-3] have been analyzing designs for mechanical systems at the nanoscale (1 to 100 nm) that could in principle be manufactured. We discuss herein some of the design optimization, and dynamics simulations, performed on these devices along with their corresponding results and analysis.

NANOSCALE MACHINES

At the nanoscale one can no longer think of the material as a continuum whose properties change continuously as it is cut and shaped. Rather one has to consider that it is formed from discrete atoms. Thus at the nanoscale one has a supermolecule rather than a finely divided solid. This requires one to analyze how the elements of normal macromaterials design change as the scale is reduced from mm to micron to nanometers. The differences are dramatic.

For a macrosystem a long alkane molecule or a flake of MoS_2 might serve as a lubricant, but for a system built on the scale of nm, such molecules may act as dirt that would clog and disable the nanosystem. Similarly the scaling of vibrations, electrical forces, thermal expansion, magnetic interactions, and surface tension with size can lead to phenomena very different at the scale of atoms than at the macroscale. Indeed in Nanosystems[2], Drexler considered many aspects of how the scaling of familiar macroscopic concepts changes as one goes to the nanoscale.

In order to make progress here it is necessary to consider specific nanoscale supermolecular systems which could in principle be manufactured. Without worrying about the practical aspects of synthesizing the system, can one make supermolecular assemblies that would function as useful nanoscale machines? Can one design synthetic procedures that could in principle manufacture these machines? Again it is not essential to have a practical cost-effective design. It is sufficient to demonstrate the existence theorem, that sequences of steps involving real atoms with the properties of real molecules can indeed lead to processes that would manufacture nanoscale machines with useful functionality. If the goal is achievable, then we can focus on ways to achieve the design in a practical fashion.

> *"Our ability to model molecular machines (systems and devices) - of specific kinds, designed in part for ease of modeling - has far outrun our ability to make them. Design calculations and computational experiments enable the theoretical study of these devices, independent of the technologies needed to implement them."*

Planetary Gear

Number of atoms: 3557

Molecular Weight: 51009.8439813

Molecular Volume: 33458.272

Along these lines Merkle and Drexler have designed a nanoscale planetary gear with eight moving parts (see Fig. 1) using molecular mechanics force fields such as DREIDING to ensure that the bond lengths, bond angles, van der Waals distances, and strain energies are reasonable. Based on their design we carried out and reported at the 1995 Nanotechnology conferences a molecular dynamics simulation to test the properties of this gear. Using somewhat extreme conditions (rapid accelerations and high frequency rotations), we found significant instabilities in this design. In response Merkle and Drexler [6]improved their designs to achieve much more stable motions.

Number of atoms: 4235

Exact Molecular Weight: 72491.9465915

Molecular Volume: 47585.964

Observed Results

We find that this new design is more stable; however, it also has too much slip. That is the actual distances between the atoms for the specific set of atoms in this supermolecular system leads to spacings between the gear teeth and the opposing gear faces that are easily stripped. The problem is that a careful balance must be achieved between:

- having the gear and race atoms so close that short range repulsive van der Waals interactions causes the gear to freeze, and
- having them too far that the gear tooth can slip past the gear tooth of the race.

It is possible to balance these in the static system, but the dynamical motions lead to fluctuations (vibrations) that allow slippage. This is not a calamity and does not mean that nanotechnology will not work. It means that we have a lot of hard work to do in designing prototype systems and testing them.

The problem here is the discreteness of the atoms. A macroscopic system is treated as a continuum. To make a gear one can design the shape of the gear teeth and the shape of the opposing teeth of the face to have exactly the same spacing as the gears turn. It is then only necessary to design the casting or injection molding system with appropriate tolerances to achieve good performance. The discreteness of the atoms plays no role when these tolerances are 0.0001 inch = 2.54 10^{-3} urn.

Neon Pump

Number of atoms: 6165

Molecular Weight: 88190.8126747

Molecular Volume: 63984.232

More recently Drexler and Merkle[3] have designed a nanomachine (Fig. 3) which in one mode could serve as a pump for Ne and in another could serve to convert Ne pressure to rotary power. We investigate this system using accurate force fields and molecular dynamics simulations to test and refine the concepts.

Static Analysis

The initial tests performed on the Neon Pump involve a 0 K simulation, using the Universal Force Fields (UFF)[7], and partial charges obtained from Gasteiger method, to establish the effect of rotor-casing, rotor-casing-Ne interactions. The following test where performed;

- Rotor static rotations with single energy calculations at each step, to determine the magnitude of forces required for inducing motion on the rotor and to determine possible empty system "frictions" along the trajectory.
- Translations along the axis of rotation where performed to evaluate energy curves during rotation and push-pull action induced by pumping action. This analysis serves as well to determine kinematic-dynamic boundaries of the system, i.e. maximum angular velocities, acceleration, etc.
- Neon atoms are introduced into the chamber of the pump and step by step 1 degree rotations of the helix followed by energy minimizatons at each step (while fixing atomic positions in the rotor and casing to avoid all atomic computation of the equations of motion, i.e. main interest on Ne trajectories).

In a sense, we can consider these simulations as being two rigid body (rotor-casing) and multiple body (rotor-casing-Ne atoms). Simulation shows a total of 245 rotation steps (5° each), for a total of 3.4 revolutions. Results clearly show higher energy due to Ne-Ne/Ne-rotor/Ne-casing interactions during their travel through the helix. The number of Ne inside the helix chamber reduces as the rotations step axis increases. Energy becomes negative since only Ne atoms are considered. The ground rules here are that a realistic FF be used and that all pieces be treated at the atomic level (but some might be semi-rigid). This will use the FF developed for nanosynthesis. Also we want to consider the effect of energy release in the chemical steps on the thermal fluctuations in these systems (which may cause displacements and vibrations).

Molecular Dynamics Simulations using Rotational Impulse Dynamics

The MD simulation initiated by a rotational impulse to the input unit. In the course of simulation this angular momentum is slowly transferred to the output unit. Over 12.0 picoseconds simulation, the input unit completes one full cycle and the output unit completes almost a half cycle and the rail completes a quarter cycle. Over the run the initial angular kinetic energy partially dissipates through conversion into thermal energy via atomic collisions. We imagine that the eventual design would have the functionality of a planetary gear but might have an appearance completely different than the macroscopic systems. Thus because a gear tooth in the xy plane cannot be atomically smooth in the z direction, we may develop a Vee design so that the Vee shape of the gear tooth in the z direction nestles within a Vee notch in the race to keep stability in the z direction as the teeth contact in the xy plane. This design would make no sense for a macroscopic gear system since the gear could never be placed inside the race. However for a molecular system one could imagine that the gear is constructed and that the race is constructed all except for a last joining unit. The parts could be assembled and then the final connections on the face made to complete the design.

The preliminary tests performed on the Neon pump demonstrate an appropriate functional behavior. The structural deformations of the rotor can cause instabilities at low and high rotational frecuencies. The forced translations show that at very low perpendicular forces due to pump action total energy rises significantly and again the structure deforms.

Control of Power Input for NanoGears and Pumps

For all three devices we study, the driver unit is a rotary component. In order to mimic operational conditions, we designed various power input profiles to drive the devices. Under such conditions these devices are not represented by thermodynamic equilibrium, but rather they are driven systems represented as in steady state thermodynamic systems.

The modes of input power profiles we used are:

1. *Impulse Mode:* A rotational angular velocity is applied to the input unit at t=0.
2. *Constant Angular Velocity:* The input unit is maintained at a constant rate of rotation about the chosen axis. Unless a heat sink is supplied (Nose-Hoover thermostat or Gaussian isokinetic thermostat, or Langevin Bath) the model system heats up, and it is driven away from the steady thermodynamic state.
3. *Time Dependent Angular Velocity:* The input unit is driven by a time dependent angular velocity.

The functional forms for time dependence should be bounded functions, especially periodic functions, sine, saw tooth, etc. This mode of operation must also be maintained at steady state by an heat bath

4. *Constant Torque - Acceleration from Rest:* This is especially important to simulate the early stages of the operation. Where a constant angular acceleration rate starts system from rest.
5. *Time Dependent Angular Acceleration Mode:* This mode is similar to that of time dependent angular velocity profile imposed on the input unit. Especially periodic functions for input torque has the system to remain within stability limits.

We have implemented these driving modes into Cerius[2] program environment[8] using the Software Developers Kit[9] version 3.0. The module utilizes graphics, communication, user interface and data model subsystems of Cerius[2]. In this implementation, user may choose any one of the energy and force evaluators. It can either be the Cerius2 Open Force Field module or an independent energy and force evaluator (user developed). In our computations we used the Cerius2 OFF module. This development is an attempt to supply a nanomachine specific simulation module[10] which could be used by various nanotechnologist. Dynamics Simulation of Gears and Neon Pump are performed at various input conditions: Constant angular velocity and constant angular acceleration for both gear designs and neon pump. We have also utilized sinusoidal driving conditions:

$W(t)=W_0 \sin (wt)$

W can either be angular velocity or angular acceleration, w is the frequency of the power source. We especially investigated the temperature profile as a function of time. To see if the sytem is driven too far from equilibrium we turned off the temperature control (Under these conditions a power source exists, but no sink mechanism is turned on.) The time variation of the kinetic temperature of various systems and driving conditions are given in the following figures.

a) Kinetic Temperature of Generation 1 Planetary Gear driven by constant initial angular velocity. Initial 2000 iterations kinetic energy maintained at a constant value.

b) Kinetic Temperature of Generation 2 Planetary Gear driven by constant angular velocity. Initial 2000 iterations kinetic energy maintained at a constant value.

c) Kinetic Temperature of Generation 2 Planetary Gear driven by a sine wave angular velocity. Initial 2000 iterations kinetic energy maintained at a constant value.

d) Cold Start with a sinusoidal driving torque applied.

CONCLUDING REMARKS

Of course, we know that nanoscale machines are possible. Nature has figured out methods of synthesizing quite remarkable nanoscale machines that function reliably, are synthesized cheaply and reliably (often self correcting) from readily available materials and energy sources, and are environmentally benign (biodegradable). Indeed modern science is finding ways to manipulate these systems to manufacture modified molecules and systems with predesigned properties and functionalities. However, the complex chemistry and biology involved in such systems limits the applications and flexibility to modify these systems for dramatically different applications. Thus it is important to determine the feasibility of designing and making nanoscale machines with new functionality and characteristics.

3

Theory of Nanoelectromechanical Systems

Due to the rapid technological advances in micro and nanofabrication techniques, the mature field of microelectromechanical systems (MEMS) has now reached the nanoscale, and consequently nanoelectromechanical systems (NEMS) have become an active field of research.

In 1998 an interesting example of a NEMS was proposed by researchers at Chalmers University in Sweden. The system consists of a movable, nanosize metallic grain spatially confined in a harmonic potential between two leads (source and drain) – see figure. In the strong Coulomb blockade regime only a single excess electron at a time is allowed to occupy the grain.

When a sufficiently high bias is applied between the leads, a single electron may tunnel onto the grain, and due to the electrostatic field between the leads, the charged grain is driven towards the drain, where the electron tunnels off. Due to the harmonic potential the uncharged grain is forced back towards the source, and the process is repeated. In this manner, the grain is driven into a self-sustained steady motion, where single electrons are "shuttled" from the source to the drain. Consequently, the device is known as a charge shuttle.

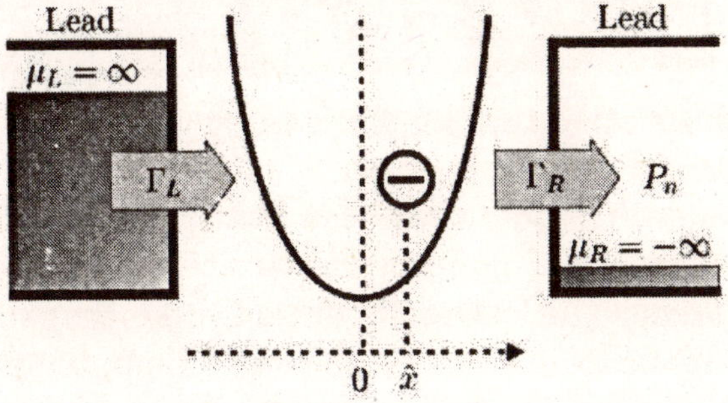

In the Theoretical Nanotechnology group we have in recent years been asking the question, what happens if the vibrations of the grain are quantized? In which aspects would such a device differ from its classical counterpart? In particular, we have studied in detail the statistics of the charge being transported through the system, the so-called full counting statistics (FCS). We have developed a systematic analytic theory for the moments (or cumulants) of the charge transport through NEMS, and with a numerical evaluation of the second cumulant (the zero-frequency current noise) we have shown that shuttle transport in the quantum regime is characterized by a very low noise-to-current ratio just as in the classical case.

Furthermore, based on a numerical evalution of the first three cumulants (average current, zero-frequency current noise, and skewness, respectively), we have shown that the device in a certain parameter regime effectively behaves as a bistable system, switching slowly between two current channels. Besides studying the quantum shuttle as described above, we have studied a related system (a vibrating quantum dot array), where the physics compared to the quantum shuttle is significantly enriched by the inclusion of additional coherently coupled states. Also in this case, the FCS reveals clear signatures of certain transport mechanisms as well as bistabilities.

WAVE–PARTICLE DUALITY

In physics and chemistry, wave–particle duality is a conceptualization that all objects in our universe exhibit properties of both waves (such as interference) and of particles (such as quantization of some of their properties). A central concept of quantum mechanics, duality addresses the inadequacy of classical concepts like "particle" and "wave" in fully describing the behaviour of quantum objects. Various interpretations of quantum mechanics attempt to explain this ostensible paradox.

The idea of duality is rooted in a debate over the nature of light and matter dating back to the 1600s, when competing theories of light were proposed by Christiaan Huygens and Isaac Newton. Through the work of Albert Einstein, Louis de Broglie and many others, current scientific theory holds that *all* particles also have a wave nature. This phenomenon has been verified not only for elementary particles, but also for compound particles like atoms and even molecules. In fact, according to standard quantum mechanics, wave-particle duality applies to all objects, even macroscopic ones; the reason we can't detect wave properties of macroscopic objects is their small wavelength.

Brief History

At the close of the 19th century, the case for atomic theory, that matter was made of particulate objects or atoms, was well established. Electricity, first thought to be a fluid, was now understood to consist of particles called electrons, as demonstrated by J.J. Thomson by his research into the work of Ernest Rutherford, who had investigated using cathode rays that an electrical charge would actually travel across a vacuum from cathode to anode. In brief, it was understood that much of nature was made of particles. At the same time, waves were well understood, together with wave phenomena such as diffraction and interference. Light was believed to be a wave, as Thomas Young's double-slit experiment and effects such as Fraunhofer diffraction had clearly demonstrated the wave-like nature of light.

But as the 20th century turned, problems had emerged. Albert Einstein's analysis of the photoelectric effect in 1905 demonstrated that light also possessed particle-like properties, and this was further confirmed with the discovery of the Compton effect in 1923. Later on, the diffraction of electrons would be predicted and experimentally confirmed, thus showing that electrons must have wave-like properties in addition to particle properties.

This confusion over particle versus wave properties was eventually resolved with the advent and establishment of quantum mechanics in the first half of the 20th century, which ultimately explained wave–particle duality. It provided a single unified theoretical framework for understanding that all matter may have characteristics associated with particles and waves, as explained below.

DEVELOPMENTAL MILESTONES

Huygens and Newton

The earliest comprehensive theory of light was advanced by Christiaan Huygens, who proposed a wave theory of light, and in particular demonstrated how waves might interfere to form a wavefront, propagating in a straight line. However, the theory had difficulties in other matters, and was soon overshadowed by Isaac Newton's corpuscular theory of light. That is, Newton proposed that light consisted of small particles, with which he could easily explain the phenomenon of reflection. With considerably more difficulty, he could also explain refraction through a lens, and the splitting of sunlight into a rainbow by a prism. Newton's particle viewpoint went essentially unchallenged for over a century.

Young, Fresnel, and Maxwell

In the early 1800s, the double-slit experiments by Young and Fresnel provided evidence for Huygens' wave theories: these experiments showed that when light is sent through a grid, a characteristic interference pattern is observed, very similar to the pattern resulting from the interference of water waves; the wavelength of light can be computed from such patterns. The wave view did not immediately displace the ray and particle view, but began to dominate scientific thinking about light in the mid 1800s, since it could explain polarization phenomena that the alternatives could not.

Maxwell, during the late-1800s, explained light as the propagation of electromagnetic waves with the Maxwell equations. These equations were verified by experiment, and Huygens' view became widely accepted.

Planck's Formula for Black Body Radiation

In 1901, Max Planck published an analysis that succeeded in reproducing the observed spectrum of light emitted by a glowing object. To accomplish this, Planck had to make an ad hoc mathematical assumption of quantized energy of the oscillators (atoms of the blackbody) that emit radiation. It was Einstein who later proposed that it is the electromagnetic radiation itself that is quantized, and not the energy of radiating atoms.

Einstein's Explanation of the Photoelectric Effect

In 1905, Albert Einstein provided an explanation of the photoelectric effect, a hitherto troubling experiment that the wave theory of light seemed incapable of explaining. He did so by postulating the existence of photons, quanta of light energy with particulate qualities.

In the photoelectric effect, it was observed that shining a light on certain metals would lead to an electric current in a circuit. Presumably, the light was knocking electrons out of the metal, causing current to flow. However, it was also observed that while a dim blue light was enough to cause a current, even the strongest, brightest red light caused no current at all. According to wave theory, the strength or amplitude of a light wave was in proportion to its brightness: a bright light should have been easily strong enough to create a large current. Yet, oddly, this was not so.

Einstein explained this conundrum by postulating that the electrons can receive energy from electromagnetic field only in discrete portions (quanta that were called photons): an amount of energy E that was related to the frequency, f of the light by

$$E = hf$$

where h is Planck's constant (6.626×10^{-34} J seconds). Only photons of a high-enough frequency, (above a certain *threshold* value) could knock an electron free. For example, photons of blue light had sufficient energy to free an electron from the metal, but photons of red light did not. More intense light above the threshold frequency could release more electrons, but no amount of light below the threshold frequency could release an electron.

Einstein was awarded the Nobel Prize in Physics in 1921 for his theory of the photoelectric effect.

Wave Behavior of Large Objects

Since the demonstrations of wave-like properties in photons and electrons, similar experiments have been conducted with neutrons and protons. Among the most famous experiments are those of Estermann and Otto Stern in 1929. Authors of similar recent experiments with atoms and molecules, described below, claim that these larger particles also act like waves.

A dramatic series of experiments emphasizing the action of gravity in relation to wave–particle duality were conducted in the 1970s using the neutron interferometer. Neutrons, one of the components of the atomic nucleus, provide much of the mass of a nucleus and thus of ordinary matter. In the neutron interferometer, they act as quantum-mechanical waves directly subject to the force of gravity. While the results were not surprising since gravity was known to act on everything, including light (see tests of general relativity and the Pound-Rebka falling photon experiment), the self-interference of the quantum mechanical wave of a massive fermion in a gravitational field had never been experimentally confirmed before.

In 1999, the diffraction of C_{60} fullerenes by researchers from the University of Vienna was reported. Fullerenes are comparatively large and massive objects, having an atomic mass of about 720u. The de Broglie wavelength is 2.5 pm, whereas the diameter of the molecule is about 1 nm, i.e. about 400 times larger. As of 2005, this is the largest object for which quantum-mechanical wave-like properties have been directly observed in far-field diffraction.

In 2003 the Vienna group also demonstrated the wave nature of tetraphenylporphyrin – a flat biodye with an extension of about 2 nm and a mass of 614u. For this demonstration they employed a near-field Talbot Lau interferometer. In the same interferometer they also found interference fringes for $C_{60}F_{48}$., a fluorinated buckyball with a mass of about 1600u, composed of 108 atoms Large molecules are already so complex that they give experimental access to some aspects of the quantum-classical interface, i.e. to certain decoherence mechanisms.

Whether objects heavier than the Planck mass (about the weight of a large bacterium) have a de Broglie wavelength is theoretically unclear and experimentally unreachable; above the Planck mass a particle's Compton wavelength would be smaller than the Planck length and its own Schwarzschild radius, a scale at which current theories of physics may break down or need to be replaced by more general ones.

Treatment in Modern Quantum Mechanics

Wave-particle duality is deeply embedded into the foundations of quantum mechanics, so well that modern practitioners rarely discuss it as such. In the formalism of the theory, all the information about a particle is encoded in its *wave function*, a complex function roughly analogous to the height of a wave at each point in space. This function evolves according to a differential equation (generically called the Schrödinger equation), and this equation gives rise to wave-like phenomena such as interference and diffraction.

The particle-like behavior is most evident due to phenomena associated with measurement in quantum mechanics. Upon measuring the location of the particle, the wave-function will randomly "collapse" to a sharply peaked function at some location, with the likelihood of any particular location equal to the squared amplitude of the wave-function there. The measurement will return a well-defined position, a property traditionally associated with particles.

Although this picture is somewhat simplified (to the non-relativistic case), it is adequate to capture the essence of current thinking on the phenomena historically called "wave-particle duality". (See also: Mathematical formulation of quantum mechanics.)

Applications

Although it is difficult to draw a line separating wave-particle duality from the rest of quantum mechanics, it is nevertheless possible to list some applications of this basic idea.

- Wave–particle duality is exploited in electron microscopy, where the small wavelengths associated with the electron can be used to view objects much smaller than what is visible using visible light.
- Similarly, neutron diffraction uses neutrons with a wavelength of about one angstrom, the typical spacing of atoms in a solid, to determine the structure of solids.

AFSHAR EXPERIMENT

The Afshar experiment is an optical experiment, which may challenge the principle of complementarity in quantum mechanics, although there is as yet no consensus on this in physics. The result of the experiment, which was first devised and carried out by Shahriar Afshar in 2001, appears to be in accordance with the standard predictions of quantum mechanics; however, it is claimed to violate the Englert-Greenberger duality relation.

The interpretation and significance of the experiment have engendered some controversy in the physics community. Afshar has published descriptions of the experiment in the American Institute of Physics and SPIE conference proceedings, and a peer-reviewed article appeared in *Foundations of Physics*. Criticisms and alternate interpretations have appeared online in blogs, at physics colloquia and academic conferences, and in arXiv e-print archives.

Overview

The principle of complementarity states that two complementary physical observables cannot both be measured for any given quantum particle without one measurement disturbing the other. An example of two complementary physical observables in quantum mechanics is the observation of the wave and particle nature of light simultaneously. The application of complementarity in this case states that we can not observe and measure the purely wave and particle like behavior of a single photon (the particle of light) at the same time. The most basic way of showing the wave like nature of light is to create interference patterns between two sources of coherent light (coherence meaning that the two sources of light have a fixed phase relationship).

When this is done, an interference pattern is created as the peaks and troughs of the two waves reinforce each other or cancel each other out. However when this is performed on with a stream of photons (each photon thus seemingly a particle), the surprising result is that the interference pattern remains. This raises the question of which way the photon went (i.e. which hole it passed through). This is highly problematic but a solution for how to think about light was formulated in the Copenhagen interpretation of quantum mechanics.

In essence this interpretation states that if it is known which way the photon goes, it is impossible to demonstrate interference. The demonstration of this is to block one of the holes, at which point the interference pattern is replaced by an apparently clear path that the photon must have taken. Some interpretations of the Afshar experiment claim that it disproves the Copenhagen interpretation, while others claim that the results are perfectly consistent with it.

Afshar's experiment uses a variant of the classic Thomas Young double-slit experiment. Such interferometer experiments typically have two "arms" or paths a photon may take. One of Afshar's assertions is that, in his experiment, it is possible to check for interference fringes of a photon stream (a measurement of the wave nature of the photons) while at the same time observing each photon's path (a measurement of the particle nature of the photons).

History

Shahriar S. Afshar's experimental work was done initially at the Institute for Radiation-Induced Mass Studies (IRIMS) in 2001 and later reproduced at Harvard University in 2003, while he was a research scholar there. The results were presented at a Harvard seminar in March 2004, and published as conference proceeding by the International Society for Optical Engineering (SPIE). The experiment was featured as the cover story in the July 24, 2004 edition of *New Scientist*. The *New Scientist* feature article itself generated many responses, including various letters to the editor that appeared in the August 7 and August 14, 2004 issues, arguing against the conclusions being drawn by Afshar, with Cramer's response. Afshar presented his work also at the American Physical Society meeting in Los Angeles, in late March of 2005.

Afshar claims that his experiment invalidates the complementarity principle and has far-reaching implications for the understanding of quantum mechanics, challenging the Copenhagen interpretation. According to John G. Cramer, Afshar's results support Cramer's own transactional interpretation of quantum mechanics and challenges the many-worlds interpretation of quantum mechanics.

Experimental setup and Afshar's interpretation

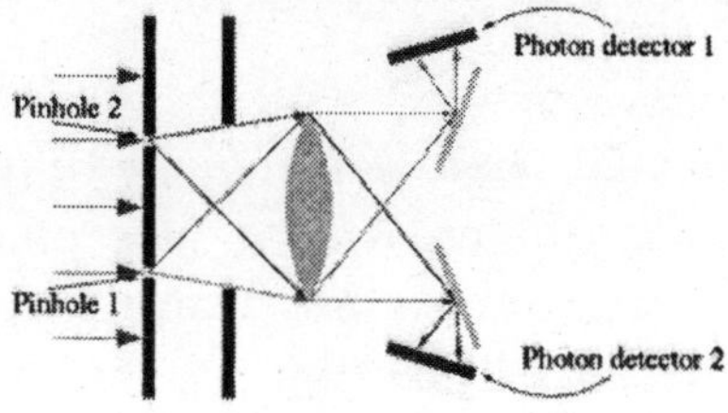

Fig.1 Experiment without obstructing wire grid

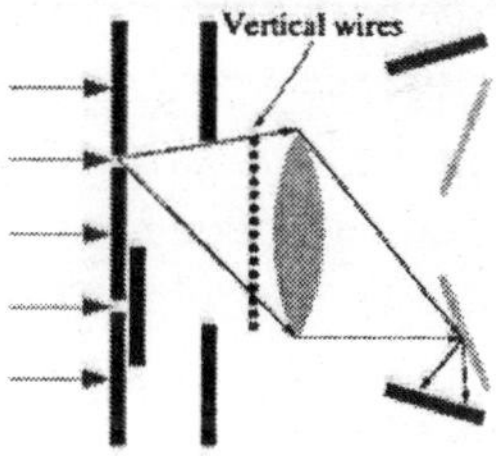

Fig. 2 Experiment with obstructing wire grid and one pinhole covered

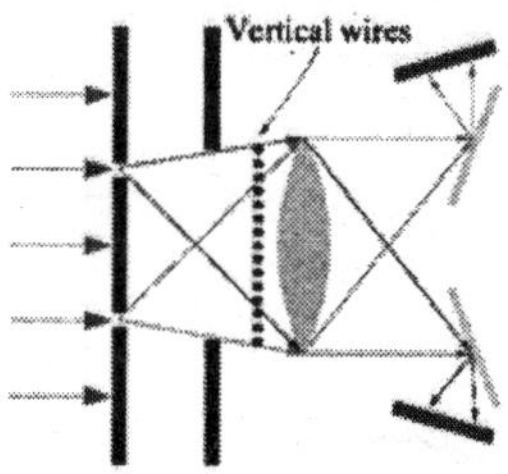

Fig. 3 Experiment with obstructing wire grid and both pinholes open

The experiment uses a setup similar to that for the double-slit experiment. In Afshar's variant, light generated by a laser passes through two closely spaced *circular* pinholes (not slits). After the dual pinholes, a lens refocuses the light so that the image of each pinhole is received by a separate photo-detector. In this setup, Afshar argues that a photon that goes through pinhole number one impinges only on detector number one, and similarly, if it goes through pinhole two. Therefore according to Afshar, if observed at the image plane, the setup is such that the light behaves as a stream of particles and can be assigned to a particular pinhole.

When the light acts as a wave, because of interference one can observe that there are regions that the photons avoid, called *dark fringes*. Afshar now places a grid of thin wires just before the lens. These wires are placed in previously measured positions of the dark fringes of an interference pattern which is produced by the dual pinhole setup when observed directly. If one of the pinholes is blocked, the interference pattern can no longer be formed, and some of the light will be blocked by the wires. Consequently, one would expect that the image quality is reduced, as is indeed observed by Afshar. Afshar then claims that he can check for the wave characteristics of the light in the same experiment, by the presence of the grid. At this point, Afshar compares the results of what is seen at the photo-detectors when one pinhole is closed with what is seen at the photo-detectors when both pinholes are open. When one pinhole is closed, the grid of wires causes some diffraction in the light, and blocks a certain amount of light received by the corresponding photo-detector. When both pinholes were open, however, the effect of the wires is minimized, so that the results are comparable to the case in which there are no wires placed in front of the lens. Afshar asserts this experiment has also been conducted with single photons and the results are identical to the high flux experiment, although these results were not available at the time of the talk at Harvard.

Afshar's conclusion is that the light exhibits a wave-like behavior when going through the wires, since the light goes through the spaces between the wires when both slits were open, but also exhibits a particle-like behavior after going through the lens, with photons going to a given photo-detector. Afshar argues that this behavior contradicts the principle of complementarity since it shows both complementary wave and particle characteristics in the same experiment for the same photons.

SPECIFIC CRITIQUES

A number of scientists have published criticisms of Afshar's interpretation of his results. While united in their rejection of Afshar, in many cases they explicitly disagree among themselves as to why he is wrong.

Some researchers claim that, while the fringe visibility is high, no which-way information ever exists: Ruth Kastner, Committee on the History and Philosophy of Science, University of Maryland, College Park.

Kastner's criticism, published in a peer-reviewed paper, proceeds by setting up a gedanken experiment and applying Afshar's logic to it to expose its flaw. She proposes that Afshar's experiment is equivalent to preparing an electron in a spin-up state and then measuring its sideways spin. This does not imply that one has found out the up-down spin state and the sideways spin state of any electron simultaneously. Applied to Afshar's experiment: "Nevertheless, even with the grid removed, since the photon is prepared in a superposition *S*, the measurement at the final screen at t_2 *never really is* a 'which-way' measurement (the term traditionally attached to the slit-basis observable), because it cannot tell us 'which slit the photon actually went through.' In addition she underscores her conclusion with an analysis of the Afshar setup within the light of the transactional interpretation of quantum mechanics. Reitzner performed numerical simulations, published in a preprint, of Afshar's arrangement and obtained the same results that Afshar obtained experimentally. From this he argues that the photons exhibit wave behavior, including high fringe visibility but no which-way information, up to the point they hit the detector: "In other words the two-peaked distribution is an interference pattern and the photon behaves as a wave and exhibits no particle properties until it hits the plate. As a result a which-way information can never be obtained in this way."

Other researchers agree that the fringe visibility is high and that the which-way information is not simultaneously measured, but they believe that the which-way information does exist under some circumstances.

Unruh, who has published his objections on the web pages of his university, is probably the most prominent critic of Afshar's interpretation. He, like Kastner, proceeds by setting up an arrangement that he feels is equivalent but simpler. The size of the effect is larger so that it is easier to see the flaw in the logic. In Unruh's view that flaw is, in the case that an obstacle exists at the position of the dark fringes, "drawing the inference that IF the particle was detected in detector 1, THEN it must have come from path 1. Similarly, IF it were detected in detector 2, then it came from path 2." In other words, he accepts the existence of an interference pattern but rejects the existence of which-way information when Afshar puts in the wire grid.

Qureshi, in his preprint, considers like Kastner a spin system as a simplified but equivalent experiment. Like Kastner he also argues that there is no which-way information, but for a very different reason. While Kastner says there is never any which-way information as long as both slits are open, Qureshi argues that there is such information when the photon passes the slits, but it is erased when the two parts of the wave overlap, irrespective of whether or not the interference pattern is actually detected. Another group does not question the which-way information, but rather contends that the measured fringe visibility is actually quite low:

Motl's criticism, published in his blog, is based on an analysis of Afshar's actual setup, instead of proposing an equivalent experiment like Unruh and Kastner. In contrast to Unruh and Kastner, he believes that which-way information always exists, but argues that the measured contrast of the interference pattern is actually very low: "Because this signal (disruption) from the second, middle picture is small (equivalently, it only affects a very small portion of the photons), the contrast V is also very small, and goes to zero for infinitely thin wires." He also argues that the experiment can be understood with classical electrodynamics and has "nothing to do with quantum mechanics".

- Aurelien Drezet, University of Graz Institut of experimental physics, Austria, Drezet's preprint argues that the classical concept of a "path" leads to much confusion in this context, but "The real problem in Afshar's interpretation comes from the fact that the interference pattern is not actually completely recorded." The argument is similar to that of Motl, that the observed visibility of the fringes is actually very small. Another way he looks at the situation is that the photons used to measure the fringes are not the same photons that are used to measure the path. The experimental setup he analyzes is only a "slightly modified version" of the one used by Afshar.

In his preprint, Steuernagel makes a quantitative analysis of the various transmitted, refracted, and reflected modes in a setup that differs only slightly from

Afshar's. He concludes that the Englert-Greenberger duality relation is strictly satisfied, and in particular that the fringe visibility for thin wires is small. Like some of the other critics, he emphasizes that inferring an interference pattern is not the same as measuring one: "Finally, the greatest weakness in the analysis given by Afshar is the *inference* that an interference pattern must be present."

ARAGO SPOT

In optics, an Arago spot is a bright point which appears at the center of the shadow of a circular object in light from a point source. It is of considerable interest due to the historical part it played in the wave theory of light. In 1818 Siméon Poisson deduced from Augustin Fresnel's theory the necessity of a bright spot at the centre of the shadow of a circular opaque obstacle. With his counterintuitive result Poisson hoped to disprove the wave theory; however Dominique Arago experimentally verifed the prediction and today the demonstration goes by the name "Poisson's (or Arago's) spot." Since the spot occurs within the geometrical shadow, no particle theory of light could account for it, and its discovery in fact provided weighty evidence for the wave nature of light, much to Poisson's chagrin.

The existence of the spot had previously been observed in 1723 by Giacomo F. Maraldi, but the work had been largely unrecognized.

The presence of Arago's spot can be easily understood. When light shines on a circular obstacle, Huygens' principle says that every point along the circumference acts as a new point source of light. But clearly, the light coming from each of these points to the center of the shadow travels exactly the same distance, and so all the light arrives in phase and constructively interferes. But this argument holds inside the object's shadow, where geometric optics and particle theories of light predict there should be no light at all.

Notice also that the above argument works specifically for a circular obstacle. Many objects have a central bright peak in their diffraction pattern, but few of those display that peak inside their shadow. The ring pattern of a Fresnel zone plate is one such arrangement.

The New Worlds Imager, a proposal to directly image extrasolar planets, relies on Arago's spot to isolate the reflected light of a planet from the sun it orbits.

HANBURY-BROWN AND TWISS EFFECT

The Hanbury Brown and Twiss effect (HBT) is any of a variety of correlation and anti-correlation effects in the intensities received by two detectors from a beam of particles. HBT effects can generally be attributed to the dual wave-particle nature of the beam, and the results of a given experiment depend on whether the beam is composed of fermions or bosons. Devices which utilize the effect are commonly called intensity interferometers and were originally used in astronomy, although they are also heavily used in the field of quantum optics.

History

In 1956, Robert Hanbury Brown and Richard Q. Twiss published *A test of a new type of stellar interferometer on Sirius*, in which two photomultiplier tubes (PMTs), separated by about 6 meters, were aimed at the star Sirius. Light was collected into the PMTs using mirrors from searchlights. An interference effect was observed between the two intensities, revealing a positive correlation between the two signals, despite the fact that no phase information was collected. Hanbury Brown and Twiss used the interference signal to determine the apparent angular size of Sirius, claiming excellent resolution.

In 1959, in order to discover the ñ^0 resonance, Goldhaber etc, [Phys.Rev.Lett.3,181(1959)] performed an experiment in Berkeley and found an unexpected angular correlation among identical pions! From then on, the HBT technique started to be used in the heavy-ion community to determine the space-time dimensions of the particle emission source for heavy ion collisions. .

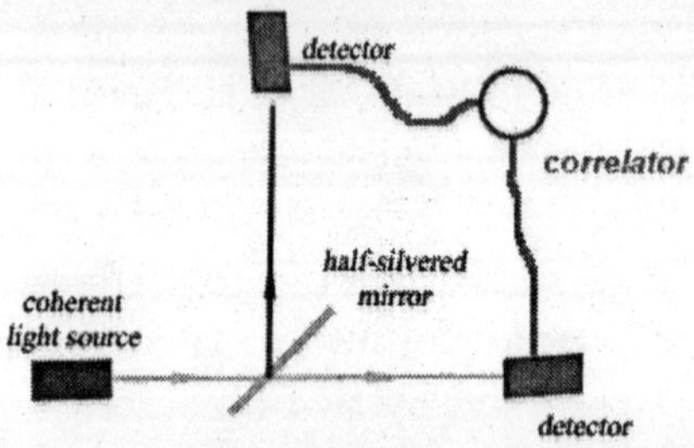

An example of an intensity interferometer that would observe no correlation, if the light source is a coherent laser beam, and positive correlations if the light source is a filtered one-mode thermal radiation. The theoretical explanation of the difference between the correlations of photon pairs in thermal and in laser beams was first given by Roy J. Glauber, who was awarded one half of the 2005 Nobel Prize in Physics "for his contribution to the quantum theory of optical coherence".

This result met with much skepticism in the physics community. Although intensity interferometry had been widely used in radio astronomy where Maxwell's equations are valid, at optical wavelengths the light would be quantised into a relatively small number of photons. Many physicists worried that the correlation was inconsistent with the laws of thermodynamics. Some even claimed that the effect violated the uncertainty principle.

Hanbury Brown and Twiss resolved the dispute in a neat series of papers which demonstrated first that wave transmission in quantum optics had exactly the same mathematical form as Maxwell's equations albeit with an additional noise term due to quantisation at the detector, and secondly that according to Maxwell's equations intensity interferometry should work. Others, such as Edward Mills Purcell immediately supported the technique, pointing out that the clumping of bosons was simply a manifestation of an effect already known in statistical mechanics. After a number of experiments, the whole physics community agreed that the observed effect was real.

The original experiment used the fact that two bosons tend to arrive at two separate detectors at the same time. Morgan and Mandel used a thermal photon source to create a dim beam of photons and observed the tendency of the photons to arrive at the same time on a single detector. Both of these effects used the wave nature of light to create a correlation in arrival time - if a single photon beam is split into two beams, then the particle nature of light requires that each photon is only observed at a single detector, and so an anti-correlation was observed in 1986. Finally, bosons have a tendency to clump together, but due to the Pauli exclusion principle, fermions tend to spread apart, and so when the Morgan and Mandel experiment is performed on electrons, an anti-correlation in arrival times was observed for the first time in 1999. All of these are considered HBT like effects.

SCATTERING THEORY

In mathematics and physics, scattering theory is a framework for studying and understanding the scattering of waves and particles. Prosaically, wave scattering corresponds to the collision and scattering of a wave with some material object, for instance sunlight scattered by rain drops to form a rainbow. Examples of particle scattering includes the motion of billiard balls or the Rutherford scattering of alpha particles by gold nuclei. More precisely, scattering consists of the study of how solutions of partial differential equations, propagating freely "in the distant past", come together and interact with one another or with a boundary condition, and then propagate way "to the distant future".

The direct scattering problem is the problem determining the distribution of scattered radiation/particle flux basing on the characteristics of the scatterer.

The inverse scattering problem is the problem of determining the characteristics of an object (its shape, internal constitution, etc.) from measurement data of radiation or particles scattered from the object.

Since its early statement for radiolocation, the problem has found vast number of applications, such as echolocation, geophysical survey, nondestructive testing, medical imaging and quantum field theory, to name just a few.

In Theoretical Physics

In mathematical physics, scattering theory is a framework for studying and understanding the interaction or scattering of solutions to partial differential equations. In acoustics, the differential equation is the wave equation, and scattering studies how its solutions, the sound waves, scatter from solid objects or propagate through non-uniform media (such as sound waves, in sea water, coming from a submarine). In the case of classical electrodynamics, the differential equation is again the wave equation, and the scattering of light or radio waves is studied. In quantum mechanics and particle physics, the equations are those of quantum electrodynamics QED, quantum chromodynamics QCD and the Standard Model, the solutions of which correspond to fundamental particles. In quantum chemistry, the solutions correspond to atoms and molecules, governed by the Schroedinger equation.

ELASTIC AND INELASTIC SCATTERING

The example of scattering in quantum chemistry is particularly instructive, as the theory is reasonably complex while still having a good foundation on which to build an intuitive understanding. When two atoms are scattered off one another, one can understand them as being the bound state solutions of some differential equation. Thus, for example, the hydrogen atom corresponds to a solution to the Schroedinger equation with a negative inverse-power (i.e., attractive Coulombic) central potential. The scattering of two hydrogen atoms will disturb the state of each atom, resulting in one or both becoming excited, or even ionized. Thus, collisions can be either elastic (the internal quantum states of the particles are not changed) or inelastic (the internal quantum states of the particles are changed). From the experimental viewpoint the observable quantity is the cross section. From the theoretical viewpoint the key quantity is the S matrix.

Topics in Physics

According to the optics classification of the Optical Society of America this field consists of the following topics:

- Aerosol and cloud effects
- Atmospheric scattering
- Backscattering
- Diffusion

- Extinction
- Index measurements
- Inverse scattering
- Linewidth
- Long-wave scattering
- Mie theory
- Multiple scattering
- Scattering measurements
- Brillouin scattering
- Molecular scattering
- Particle scattering
- Raman scattering
- Rayleigh Scattering
- Scattering from rough surfaces
- Stimulated scattering
- Stimulated Brillouin scattering
- Stimulated Raman scattering
- Scintillation
- Turbid media

The Mathematical Framework

In mathematics, scattering theory deals with a more abstract formulation of the same set of concepts. For example, if a differential equation is known to have some simple, localized solutions, and the solutions are a function of a single parameter, that parameter can take the conceptual role of time. One then asks what might happen if two such solutions are set up far away from each other, in the "distant past", and are made to move towards each other, interact (under the constraint of the differential equation) and then move apart in the "future". The scattering matrix then pairs solutions in the "distant past" to those in the "distant future".

Solutions to differential equations are often posed on manifolds. Frequently, the means to the solution requires the study of the spectrum of an operator on the manifold. As a result, the solutions often have a spectrum that can be identified with a Hilbert space, and scattering is described by a certain map, the S matrix, on Hilbert spaces. Spaces with a discrete spectrum correspond to bound states in quantum mechanics, while a continuous spectrum is associated with scattering states. The study of inelastic scattering then asks how discrete and continuous spectra are mixed together.

An important, notable development is the inverse scattering transform, central to the solution of many exactly solvable models.

DARK MATTER

In astrophysics and cosmology, dark matter refers to hypothetical matter of unknown composition that does not emit or reflect enough electromagnetic radiation to be observed directly, but whose presence can be inferred from gravitational effects on visible matter. According to present observations of structures larger than galaxy-sized as well as Big Bang cosmology, dark matter accounts for the vast majority of mass in the observable universe.

Fritz Zwicky used it for the first time to declare the observed phenomena consistent with dark matter observations as the rotational speeds of galaxies and orbital velocities of galaxies in clusters, gravitational lensing of background objects by galaxy clusters such as the Bullet cluster, and the temperature distribution of hot gas in galaxies and clusters of galaxies. Dark matter also plays a central role in structure formation and galaxy evolution, and has measurable effects on the anisotropy of the cosmic microwave background. All these lines of evidence suggest that galaxies, clusters of galaxies, and the universe as a whole contain far more matter than that which interacts with electromagnetic radiation: the remainder is called the "dark matter component".

The composition of dark matter is unknown, but may include ordinary and heavy neutrinos, recently postulated elementary particles such as WIMPs and axions, astronomical bodies such as dwarf stars and planets (collectively called MACHOs), and clouds of nonluminous gas. Current evidence favors models in which the primary component of dark matter is new elementary particles, collectively called non-baryonic dark matter.

The dark matter component has vastly more mass than the "visible" component of the universe. At present, the density of ordinary baryons and radiation in the universe is estimated to be equivalent to about one hydrogen atom per cubic metre of space. Only about 4% of the total energy density in the universe (as inferred from gravitational effects) can be seen directly. About 22% is thought to be composed of dark matter. The remaining 74% is thought to consist of dark energy, an even stranger component, distributed diffusely in space. Some hard-to-detect baryonic matter makes a contribution to dark matter, but constitutes only a small portion. Determining the nature of this missing mass is one of the most important problems in modern cosmology and particle physics. It has been noted that the names "dark matter" and "dark energy" serve mainly as expressions of our ignorance, much as the marking of early maps with "terra incognita".

Observational Evidence

The first to provide evidence and infer the existence of a phenomenon that has come to be called "dark matter" was Swiss astrophysicist Fritz Zwicky, of the California Institute of Technology (Caltech) in 1933. He applied the virial theorem to the Coma cluster of galaxies and obtained evidence of unseen mass. Zwicky estimated the cluster's total mass based on the motions of galaxies near its edge. When he compared this mass estimate to one based on the number of galaxies and total brightness of the cluster, he found that there was about 400 times more mass than expected.

The gravity of the visible galaxies in the cluster would be far too small for such fast orbits, so something extra was required. This is known as the "missing mass problem". Based on these conclusions, Zwicky inferred that there must be some non-visible form of matter which would provide enough of the mass and gravity to hold the cluster together.

Much of the evidence for dark matter comes from the study of the motions of galaxies. Many of these appear to be fairly uniform, so by the virial theorem the total kinetic energy should be half the total gravitational binding energy of the galaxies.

Experimentally, however, the total kinetic energy is found to be much greater: in particular, assuming the gravitational mass is due to only the visible matter of the galaxy, stars far from the center of galaxies have much higher velocities than predicted by the virial theorem.

Galactic rotation curves, which illustrate the velocity of rotation versus the distance from the galactic center, cannot be explained by only the visible matter. Assuming that the visible material makes up only a small part of the cluster is the most straightforward way of accounting for this. Galaxies show signs of being composed largely of a roughly spherically symmetric, centrally concentrated halo of dark matter with the visible matter concentrated in a disc at the center. Low surface brightness dwarf galaxies are important sources of information for studying dark matter, as they have an uncommonly low ratio of visible matter to dark matter, and have few bright stars at the center which impair observations of the rotation curve of outlying stars.

According to results published in August 2006, dark matter has been observed separate from ordinary matter through measurements of the Bullet Cluster, actually two nearby clusters of galaxies that collided about 150 million years ago. Researchers analyzed the effects of gravitational lensing to determine total mass distribution in the pair and compared that to X-ray maps of hot gases, thought to constitute the large majority of ordinary matter in the clusters. The hot gases interacted during the collision and remain closer to the center. The individual galaxies and the dark matter did not interact and are farther from the center.

Dr. Myungkook James Jee and his colleagues announced on May 15, 2007 the discovery of a wispy ring of dark matter 2.6 million light-years wide that envelopes CL0024+17, a huge cluster of galaxies about 5 billion light-years away .

Their observation of the dark matter was by way of its gravitational lensing effect on light coming from behind the galaxy cluster as seen by the now broken Hubble Space Telescope's Advanced Camera for Surveys (ACS). Richard Massey, Cal Tech publisher of a dark matter map for a half million galaxies, notes that this announcement only comes from one instrument and that "the signal is very weak. Some people are not yet convinced it's more than an artifact." Confirming studies may need to wait until the James Webb Space Telescope is launched in 2013 unless the Hubble's ACS is repaired by a shuttle mission.

Galactic Rotation Curves

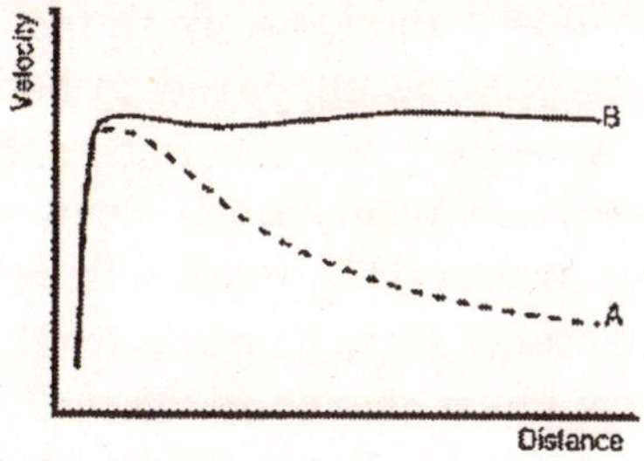

Rotation curve of a typical spiral galaxy: predicted (A) and observed (B). Dark matter can explain the velocity curve having a "flat" appearance out to a large radius

For nearly 40 years after Zwicky's initial observations, no other corroborating observations indicated that the mass to light ratio was anything other than unity (a high mass-to-light ratio indicates the presence of dark matter). Then, in the late 1960s and early 1970s, Vera Rubin, a young astronomer at the Department of Terrestrial Magnetism at the Carnegie Institution of Washington presented findings based on a new sensitive spectrograph that could measure the velocity curve of edge-on spiral galaxies to a greater degree of accuracy than had ever before been achieved.

Together with fellow staff-member Kent Ford, Rubin announced at a 1975 meeting of the American Astronomical Society the astonishing discovery that most stars in spiral galaxies orbit at roughly the same speed, which implied that their mass densities were uniform well beyond the locations with most of the stars (the galactic bulge). This result suggests that either Newtonian gravity does not apply universally or that, conservatively, upwards of 50% of the mass of galaxies was contained in the relatively dark galactic halo.

Met with skepticism, Rubin insisted that the observations were correct. Eventually other astronomers began to corroborate her work and it soon became well-established that most galaxies were in fact dominated by "dark matter"; exceptions appeared to be galaxies with mass-to-light ratios close to that of stars. Subsequent to this, numerous observations have been made that do indicate the presence of dark matter in various parts of the cosmos. Together with Rubin's findings for spiral galaxies and Zwicky's work on galaxy clusters, the observational evidence for dark matter has been collecting over the decades to the point that today most astrophysicists accept its existence. As a unifying concept, dark matter is one of the dominant features considered in the analysis of structures on the order of galactic scale and larger.

VELOCITY DISPERSIONS OF GALAXIES

Rubin's pioneering work has stood the test of time. Measurements of velocity curves in spiral galaxies were soon followed up with velocity dispersions of elliptical galaxies. While sometimes appearing with lower mass-to-light ratios, measurements of ellipticals still indicate a relatively high dark matter content. Likewise, measurements of the diffuse interstellar gas found at the edge of galaxies indicate not only dark matter distributions that extend beyond the visible limit of the galaxies, but also that the galaxies are virialized up to ten times their visible radii. This has the effect of pushing up the dark matter as a fraction of the total amount of gravitating matter from 50% measured by Rubin to the now accepted value of nearly 95%.

There are places where dark matter seems to be a small component or totally absent. Globular clusters show no evidence that they contain dark matter, though their orbital interactions with galaxies do show evidence for galactic dark matter.

For some time, measurements of the velocity profile of stars seemed to indicate concentration of dark matter in the disk of the Milky Way galaxy, however, now it seems that the high concentration of baryonic matter in the disk of the galaxy (especially in the interstellar medium) can account for this motion. Galaxy mass profiles are thought to look very different from the light profiles. The typical model for dark matter galaxies is a smooth, spherical distribution in virialized halos. Such would have to be the case to avoid small-scale (stellar) dynamical effects. Recent research reported in January 2006 from the University of Massachusetts, Amherst would explain the previously mysterious warp in the disk of the Milky Way by the interaction of the Large and Small Magellanic Clouds and the predicted 20 fold increase in mass of the Milky Way taking into account dark matter.

Recently (2005), astronomers from Cardiff University claim to have discovered a galaxy made almost entirely of dark matter, 50 million light years away in the Virgo Cluster, which was named VIRGOHI21. Unusually, VIRGOHI21 does not appear to contain any visible stars: it was seen with radio frequency observations of hydrogen. Based on rotation profiles, the scientists estimate that this object contains approximately 1000 times more dark matter than hydrogen and has a total mass of about 1/10th that of the Milky Way Galaxy we live in. For comparison, the Milky Way is believed to have roughly 10 times as much dark matter as ordinary matter. Models of the Big Bang and structure formation have suggested that such dark galaxies should be very common in the universe, but none have previously been detected. If the existence of this dark galaxy is confirmed, it provides strong evidence for the theory of galaxy formation and poses problems for alternative explanations of dark matter.

Missing Matter in Clusters of Galaxies

Dark matter affects galaxy clusters as well. X-ray measurements of hot intracluster gas correspond closely to Zwicky's observations of mass-to-light ratios for large clusters of nearly 10 to 1. Many of the experiments of the Chandra X-ray Observatory use this technique to independently determine the mass of clusters.

The galaxy cluster Abell 2029 is composed of thousands of galaxies enveloped in a cloud of hot gas, and an amount of dark matter equivalent to more than 10^{14} Suns. At the center of this cluster is an enormous, elliptically shaped galaxy that is thought to have been formed from the mergers of many smaller galaxies. The measured orbital velocities of galaxies within galactic clusters have been found to be consistent with dark matter observations.

Another important tool for future dark matter observations is gravitational lensing. Lensing relies on the effects of general relativity to predict masses without relying on dynamics, and so is a completely independent means of measuring the dark matter. Strong lensing, the observed distortion of background galaxies into arcs when the light passes through a gravitational lens, has been observed around a few distant clusters including Abell 1689 (pictured right). By measuring the distortion geometry, the mass of the cluster causing the phenomena can be obtained. In the dozens of cases where this has been done, the mass-to-light ratios obtained correspond to the dynamical dark matter measurements of clusters.

Perhaps more convincing, a technique has been developed over the last 10 years called weak lensing which looks at microscale distortions of galaxies observed in vast galaxy surveys due to foreground objects through statistical analyses. By examining the shear deformation of the adjacent background galaxies, astrophysicists can characterize the mean distribution of dark matter by statistical means and have found mass-to-light ratios that correspond to dark matter densities predicted by other large-scale structure measurements. The correspondence of the two gravitational lens techniques to other dark matter measurements has convinced almost all astrophysicists that dark matter actually exists as a major component of the universe's composition.

Structure Formation

Dark matter is crucial to the Big Bang model of cosmology as a component which corresponds directly to measurements of the parameters associated with Friedmann cosmology solutions to general relativity. In particular, measurements of the cosmic microwave background anisotropies correspond to a cosmology where much of the matter interacts with photons more weakly than the known forces that couple light interactions to baryonic matter. Likewise, a significant amount of non-baryonic, cold matter is necessary to explain the large-scale structure of the universe.

Observations suggest that structure formation in the universe proceeds hierarchically, with the smallest structures collapsing first and followed by galaxies and then clusters of galaxies. As the structures collapse in the evolving universe, they begin to "light up" as the baryonic matter heats up through gravitational contraction and the object approaches hydrostatic pressure balance. Ordinary baryonic matter had too high a temperature, and too much pressure left over from the big bang to collapse and form smaller structures, such as stars, via the Jeans instability. Dark matter acts as a compactor of structure. This model not only corresponds with statistical surveying of the visible structure in the universe but also corresponds precisely to the dark matter predictions of the cosmic microwave background. This *bottom up* model of structure formation requires something like cold dark matter to succeed. Large computer simulations of billions of dark matter particles have been used to confirm that the cold dark matter model of structure formation is consistent with the structures observed in the universe through galaxy surveys, such as the Sloan Digital Sky Survey and 2dF Galaxy Redshift Survey, as well as observations of the Lyman-alpha forest. These studies have been crucial in constructing the Lambda-CDM model which measures the cosmological parameters, including the fraction of the universe made up of baryons and dark matter.

Dark Matter Composition

Although dark matter was detected by its gravitational lensing in August 2006, many aspects of dark matter remain speculative. The DAMA/NaI experiment has claimed to directly detect dark matter passing through the Earth, though most scientists remain skeptical since negative results of other experiments are (almost) incompatible with the DAMA results if dark matter consists of neutralinos.

Data from a number of lines of evidence, including galaxy rotation curves, gravitational lensing, structure formation, and the fraction of baryons in clusters and the cluster abundance combined with independent evidence for the baryon density, indicate that 85-90% of the mass in the universe does not interact with the electromagnetic force. This "dark matter" is evident through its gravitational effect. Several categories of dark matter have been postulated.

- Baryonic dark matter
- Non-baryonic dark matter which is divided into three different types:

- o Hot dark matter - nonbaryonic particles that move ultrarelativistically
- o Warm dark matter - nonbaryonic particles that move relativistically
- o Cold dark matter - nonbaryonic particles that move non-relativistically

Davis *et al* wrote in 1985:

"Candidate particles can be grouped into three categories on the basis of their effect on the fluctuation spectrum (Bond et al *1983). If the dark matter is composed of abundant light particles which remain relativistic until shortly before recombination, then it may be termed "hot". The best candidate for hot dark matter is a neutrino [..]*

A second possibility is for the dark matter particles to interact more weakly than neutrinos, to be less abundant, and to have a mass of order 1eV. Such particles are termed "warm dark matter", because they have lower thermal velocities than massive neutrinos [..] there are at present few candidate particles which fit this description. Gravitinos and photinos have been suggested (Pagels and Primack 1982; Bond, Szalay and Turner 1982) [..]

Any particles which became nonrelativistic very early, and so were able to diffuse a negligible distance, are termed "cold" dark matter (CDM). There are many candidates for CDM including supersymmetric particles "

Hot dark matter consists of particles that travel with relativistic velocities. One kind of hot dark matter is known, the neutrino. Neutrinos have a very small mass, do not interact via either the electromagnetic or the strong nuclear force and are therefore very difficult to detect. This is what makes them appealing as dark matter. However, bounds on neutrinos indicate that ordinary neutrinos make only a small contribution to the density of dark matter.

Hot dark matter cannot explain how individual galaxies formed from the Big Bang. The microwave background radiation as measured by the COBE and WMAP satellites, while incredibly smooth, indicates that matter has clumped on very small scales. Fast moving particles, however, cannot clump together on such small scales and, in fact, suppress the clumping of other matter. Hot dark matter, while it certainly exists in our universe in the form of neutrinos, is therefore only part of the story.

The Concordance Model requires that, to explain structure in the universe, it is necessary to invoke cold (non-relativistic) dark matter. Large masses, like galaxy-sized black holes can be ruled out on the basis of gravitational lensing data. Possibilities involving normal baryonic matter include brown dwarfs or perhaps small, dense chunks of heavy elements; such objects are known as massive compact halo objects, or "MACHOs". However, studies of big bang nucleosynthesis have convinced most scientists that baryonic matter such as MACHOs cannot be more than a small fraction of the total dark matter.

At present, the most common view is that dark matter is primarily non-baryonic, made of one or more elementary particles other than the usual electrons, protons, neutrons, and known neutrinos. The most commonly proposed particles are axions, sterile neutrinos, and WIMPs (Weakly Interacting Massive Particles, including neutralinos). None of these are part of the standard model of particle physics, but they can arise in extensions to the standard model. Many supersymmetric models naturally give rise to stable WIMPs in the form of neutralinos. Heavy, sterile neutrinos exist in extensions to the standard model that explain the small neutrino mass through the seesaw mechanism.

Experimental searches for these dark matter candidates have been conducted and are ongoing. These efforts can be divided into two broad classes: direct detection, in which the dark matter particles are observed in a detector; and indirect detection, which looks for the products of dark matter annihilations. Dark matter detection experiments have ruled out some WIMP and axion models. There are also several experiments claiming positive evidence for dark matter detection, such as DAMA/NaI, PVLAS, and EGRET, but these are so far unconfirmed and difficult to reconcile with the negative results of other experiments. Several searches for dark matter are currently underway, including the Cryogenic Dark Matter Search in the Soudan mine and the XENON experiment at Gran Sasso, and many new technologies are under development, such as the ArDM experiment.

In research due to be fully published in spring 2006, researchers from the University of Cambridge Institute of Astronomy claim to have calculated that dark matter only comes in clumps larger than about 1,000 light-years across, implying an average speed of dark matter particles of 9 km/s, a density of 20 amu/cm^3, and temperature of 10,000 kelvins.

ALTERNATIVE EXPLANATIONS

Modifications of Gravity

A proposed alternative to physical dark matter particles has been to suppose that the observed inconsistencies are due to an incomplete understanding of gravitation. To explain the observations, the gravitational force has to become stronger than the Newtonian approximation at great distances or in weak fields. One of the proposed models is Modified Newtonian Dynamics (MOND), which corrects Newton's laws at small acceleration.

However, constructing a relativistic MOND theory has been troublesome, and it is not clear how the theory can be reconciled with gravitational lensing measurements of the deflection of light around galaxies. The leading relativistic MOND theory, proposed by Jacob Bekenstein in 2004 is called TeVeS for Tensor-Vector-Scalar and solves many of the problems of earlier attempts. A theory of modified gravity (MOG) proposed by John W. Moffatt, based upon the Nonsymmetric Gravitational Theory (NGT), is also an alternative to dark matter.

In August 2006, a study of colliding galaxy clusters claimed to show that even in a modified gravity hypothesis, the majority of the mass must be some form of dark matter by demonstrating that when regular matter is "swept away" from a cluster, the gravitational effects of dark matter (which is thought to be non-interacting aside from its gravitational effect) remain. A study claims that TeVeS may be able to produce the observed effect, but this still requires the majority of the mass to be in the form of dark matter, possibly in the form of ordinary neutrinos. Also Nonsymmetric Gravitational Theory has been claimed to qualitatively fit the observations without needing exotic dark matter.

QUANTUM MECHANICAL EXPLANATIONS

In another class of theories one attempts to reconcile gravitation with quantum mechanics and obtains corrections to the conventional gravitational interaction. In scalar-tensor theories, scalar fields like the Higgs field couples to the curvature given through the Riemann tensor or its traces. In many of such theories, the scalar field equals the inflaton field, which is needed to explain the inflation of the universe after the Big Bang, as the dominating factor of the quintessence or Dark Energy. Using an approach based on the exact renormalization group, M. Reuter and H. Weyer have shown that Newton's constant and the cosmological constant can be scalar functions on spacetime if one associates renormalization scales to the points of spacetime.

In Laurent Nottale's scale relativity theory, spacetime is continuous but not differentiable, leading to the appearance of a gravitational Schrödinger equation. As a result, quantization effects appear at large scale. This makes it possible to predict correctly the large scale structures of the universe without the need for the dark matter hypothesis.

Dark Matter in Popular Culture

Mentions of dark matter occur in some video games and other works of fiction. In such cases, it is usually attributed extraordinary physical or magical properties. Such descriptions are often inconsistent with the properties of dark matter proposed in physics and cosmology.

STERN–GERLACH EXPERIMENT

In quantum mechanics, the Stern–Gerlach experiment, named after Otto Stern and Walther Gerlach, is a celebrated experiment (performed in 1922) on the deflection of particles, often used to illustrate basic principles of quantum mechanics. It can be used to demonstrate that electrons and atoms have intrinsically quantum properties, and how measurement in quantum mechanics affects the system being measured.

Basic Theory and Description

Otto Stern and Walther Gerlach devised an experiment to determine whether particles had any intrinsic angular momentum. In a classical system, such as the earth orbiting the sun, the earth has angular momentum from both its orbit around the sun and the orbit around its axis (its spin). The experiment sought to determine whether individual particles like electrons have any "spin" angular momentum. If the electron is treated like a classical dipole with two halves of charge spinning quickly, it will begin to precess in a magnetic field, because of the torque that the magnetic field exerts on the dipole.

If the particle travels in a homogeneous magnetic field, the forces exerted on opposite ends of the dipole cancel each other out and the motion of the particle is unaffected. If the experiment is conducted using electrons, an electric field of appropriate magnitude and oriented transverse to the charged particle's path is used to compensate for the tendency of any charged particle to curl in its path through a magnetic field (see cyclotron motion), and the fact that electrons are charged can safely be ignored. The Stern–Gerlach experiment can be conducted using electrically neutral particles and the same conclusion is reached, since it is designed to test angular momentum only, not any electrostatic phenomena.

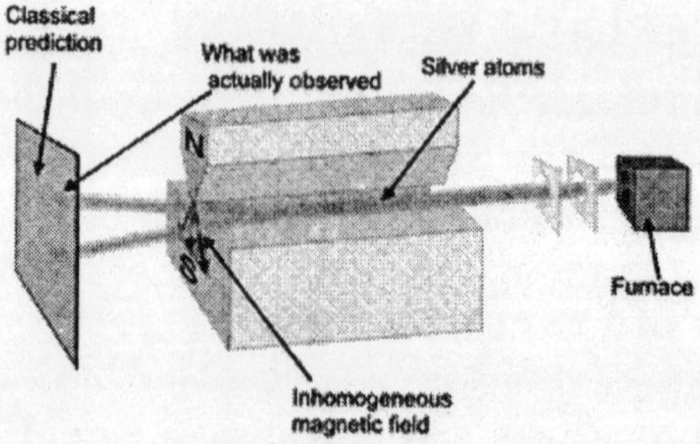

Basic elements of the Stern–Gerlach experiment.

If the particle travels through an inhomogeneous magnetic field, then the force on one end of the dipole will be slightly greater than the opposing force on the other end of the dipole. This leads to the particle being deflected in the inhomogeneous magnetic field. The direction in which the particles are deflected is typically called the "z" direction.

If the particles are classical, "spinning" particles, then the distribution of their spin angular momentum vectors is taken to be truly random and each particle would be deflected up or down by a different amount, producing an even distribution on the screen of a detector. Instead, the particles passing through the device are deflected either up or down by a specific amount. This can only mean that spin angular momentum is quantized, i.e. it can only take on discrete values. There is not a continuous distribution of possible angular momenta.

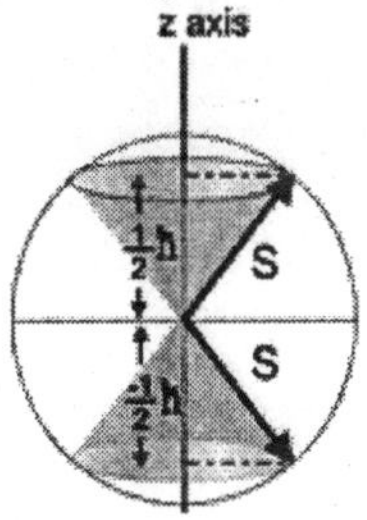

Spin values for fermions.

Sequential experiments

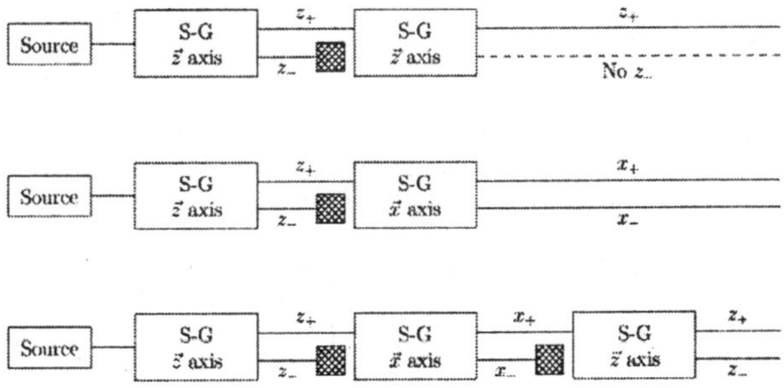

If we combine some Stern–Gerlach apparati we can clearly see that they do not act as simple selectors, but alter the states observed (as in light polarization), according to quantum mechanics laws:

History

The Stern–Gerlach experiment was performed in Frankfurt, Germany in 1922 by Otto Stern and Walther Gerlach. At the time, Stern was an assistant to Max Born at the University of Frankfurt's Institute for Theoretical Physics, and Gerlach was an assistant at the same university's Institute for Experimental Physics.

At the time of the experiment, the most prevalent model for describing the atom was the Bohr model, which described electrons as going around the positively-charged nucleus only in certain discrete atomic orbitals or energy levels. Since the electron was quantized to be only in certain positions in space, the separation into distinct orbits was referred to as space quantization.

Impact

The Stern–Gerlach experiment had one of the biggest impacts on modern physics:

- In the decade that followed, scientists showed using similar techniques, that the nucleus of some atoms also have quantized angular momentum. It is the interaction with the spin of the electron that is responsible for the hyperfine structure of the spectroscopic lines.
- In the thirties, using an extended version of the S–G apparatus, Isidor Rabi and colleagues showed that by using a varying magnetic field, one can force the magnetic momentum to go from one state to the other. The series of experiments culminated in 1937 when they discovered that state transitions

could be induced using time varying fields or RF fields. The so called Rabi oscillation is the working mechanism for the Magnetic Resonance Imaging equipment found in hospitals.

- Later Norman F. Ramsey, modified the Rabi apparatus to increase the interaction time with the field. The extreme sensitivity due to frequency of the radiation makes this very useful for keeping accurate time, and is still used today in atomic clocks.
- In the early sixties, Ramsey and Daniel Kleppner used a S–G system to produce a beam of polarized hydrogen as the source of energy for the Hydrogen Maser, which is still one of the most popular atomic clocks.
- The direct observation of the spin is the most direct proof of quantization in quantum mechanics.

DOUBLE-SLIT EXPERIMENT

The double-slit experiment generally refers to an experiment in which light is allowed to diffract through two slits, which produces fringes or wave-like interference patterns on an opposing screen.

These interference patterns will result in projected light and dark regions that correspond to where the light waves have constructively (added) and destructively (subtracted) interfered.

The experiment can also be performed with a beam of electrons or atoms, showing similar interference patterns; this is taken as evidence of the "wave-particle duality" predicted by quantum physics. Note, however, that a double-slit experiment can also be performed with water waves in a ripple tank; the explanation of the observed wave phenomena does not require quantum mechanics in any way. The phenomenon is quantum mechanical only when quantum particles—such as atoms, electrons, or photons—manifest as waves.

Importance to physics

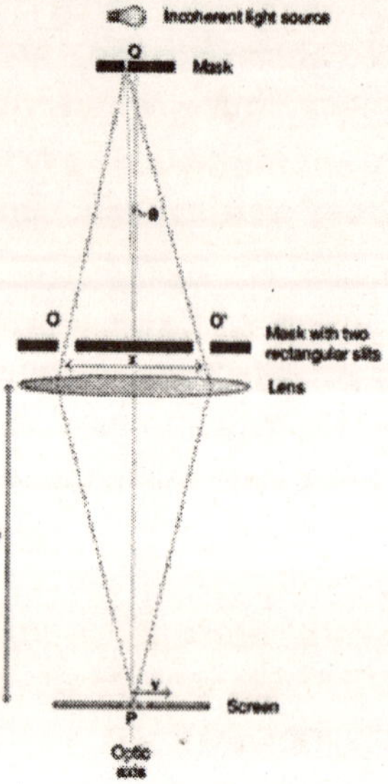

Sketch of the layout of a typical optical double-slit experiment

Although the double-slit experiment is now often referred to in the context of quantum mechanics, it is generally thought to have been first performed by the English scientist Thomas Young around the year 1800 in an attempt to resolve the question of whether light was composed of particles (Newton's "corpuscular" theory), or rather consisted of waves traveling through some ether, just as sound waves travel in air. The interference patterns observed in the experiment seemed to discredit the corpuscular theory, and the wave theory of light remained well accepted until the early 20th century, when evidence began to accumulate which seemed instead to confirm the particle theory of light.

The double-slit experiment, and its variations, then became a classic *Gedankenexperiment* (thought experiment) for its clarity in expressing the central puzzles of quantum mechanics; although in this form the experiment was not actually performed with anything other than light until 1961, when Claus Jönsson of the University of Tübingen performed it with electrons, and not until 1974 in the form of "one electron at a time", in a laboratory at the University of Milan, by researchers led by Pier Giorgio Merli, of LAMEL-CNR Bologna.

The results of the 1974 experiment were published and even made into a short film, but did not receive wide attention. The experiment was repeated in 1989 by Tonomura et al at Hitachi in Japan. Their equipment was better, reflecting 15 years of advances in electronics and a dedicated development effort by the Hitachi team. Their methodology was more precise and elegant, and their results agreed with the results of Merli's team. Although Tonomura asserted that the Italian experiment had not detected electrons one at a time—a key to demonstrating the wave-particle paradox—single electron detection is clearly visible in the photos and film taken by Merli and his group.

In September 2002, the double-slit experiment of Claus Jönsson was voted "the most beautiful experiment" by readers of *Physics World*.

Results Observed

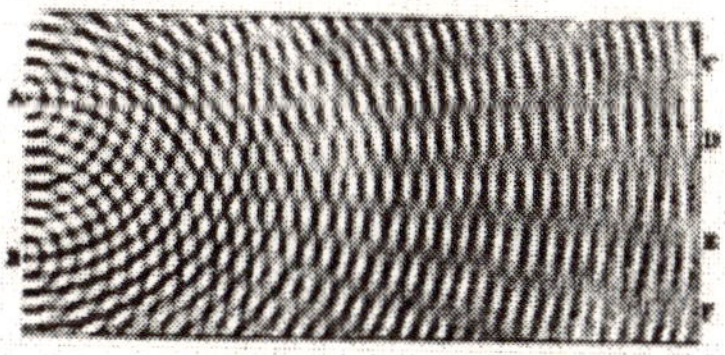

Thomas Young's sketch of two-slit diffraction, based on his observations of water waves.

The bright bands observed on the screen happen when the light has interfered constructively—where a crest of a wave meets a crest from another wave. The dark regions show destructive interference—a crest meets a trough. Constructive interference occurs when

$$\frac{n\lambda}{d} = \frac{x}{L} \quad \Leftrightarrow \quad n\lambda = \frac{xd}{L},$$

where

ë is the wavelength of the light,

d is the separation of the slits,

n is the order of maxima observed (central maximum is n=0),

x is the distance between the bands of light and the central maximum (also called fringe distance), and

L is the distance from the slits to the screen.

This is only an approximation and depends on certain conditions.

It is possible to work out the wavelength of light using this equation and the above apparatus. If *d* and *L* are known and *x* is observed, then *ë* can be easily calculated.

A detailed treatment of the mathematics of double-slit interference in the context of quantum mechanics is given in the article on Englert-Greenberger duality.

Shape of Interference Fringes

The theoretical shapes of the interference fringes observed in Young's double slit experiment are straight lines which is easily proved.

In case two pinholes are used instead of slits, as in the original Young's experiment, hyperbolic fringes are observed. This is because the difference in paths travelled by the light from the two sources is a constant for a fringe which is the property of a hyperbola.

If the two sources are placed on a line perpendicular to the screen, the shape of the interference fringes is circular as the individual paths travelled by light from the two sources are always equal for a given fringe. This can be done in simpler way by placing a mirror parallel to a screen at a distance and a source of light just above the mirror. (Note the extra phase difference of ∂ due to reflection at the interface of a denser medium)

Quantum Version of Experiment

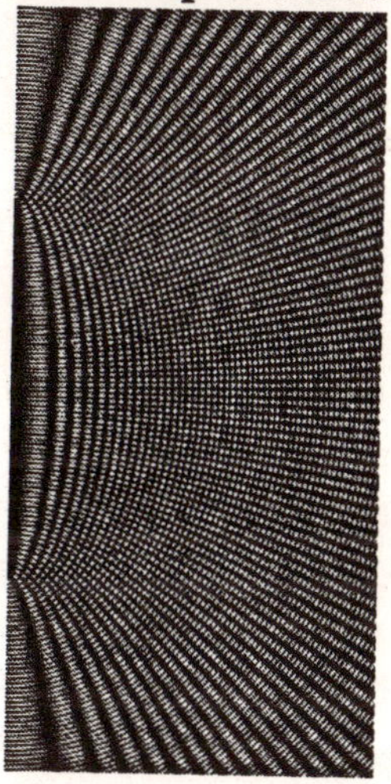

The wavefronts resulting from two pinholes.

By the 1920s, various other experiments (such as the photoelectric effect) had demonstrated that light interacts with matter only in discrete, "quantum"-sized packets called photons.

If sunlight is replaced with a light source that is capable of producing just one photon at a time, and the screen is sensitive enough to detect a single photon, Young's experiment can, in theory, be performed one photon at a time—with identical results. If either slit is covered, the individual photons hitting the screen, over time, create a pattern with a single peak. But if both slits are left open, the pattern of photons hitting the screen, over time, again becomes a series of light and dark fringes. This result seems to both confirm and contradict the wave theory. On the one hand, the interference pattern confirms that light still behaves much like a wave, even though we send it one particle at a time. On the other hand, each time a photon with a certain energy is emitted, the screen detects a photon with the same energy. Under the Copenhagen interpretation of quantum theory, an individual photon is seen as passing through both slits at once, and interfering with itself, producing the interference pattern.

A remarkable result follows from a variation of the double-slit experiment, in which detectors are placed in each of the two slits, in an attempt to determine which slit the photon passes through on its way to the screen. Placing a detector even in just one of the slits will result in the disappearance of the interference pattern.

The detection of a photon involves a physical interaction between the photon and the detector of the sort that physically changes the detector. (If nothing changed in the detector, it would not detect anything.) If two photons of the same frequency were emitted at the same time they would be coherent. If they went through two unobstructed slits then they would remain coherent and arriving at the screen at the same time but laterally displaced from each other they would exhibit interference.

However, if one or both of them were to encounter a detector, then they would fall out of step with each other, that is, they would decohere. They would then arrive at the screen at slightly different times and could not interfere because the first to arrive would have already interacted with the screen before the second got there.

If only one photon is involved, it must be detected at one or the other detector, and its continued path goes forward only from the slit where it was detected.

The Copenhagen interpretation posits the existence of probability waves which describe the likelihood of finding the particle at a given location. Until the particle is detected at any location along this probability wave, it effectively exists *at every point*.

Thus, when the particle could be passing through either of the two slits, it will actually pass through both, and so an interference pattern results. But if the particle is detected at one of the two slits, then it can no longer be passing through both—its presence must become manifested at one or the other, and so no interference pattern appears.

This is similar to the path integral formulation of quantum mechanics provided by Richard Feynman. (Feynman stresses that this is merely a mathematical description, not an attempt to describe some "real" process that we cannot see.)

In the path integral formulation, a particle such as a photon takes *every possible path* through space-time to get from point A to point B. In the double-slit experiment, point A might be the emitter, and point B the screen upon which the interference pattern appears, and a particle takes every possible path — through both slits at once— to get from A to B.

When a detector is placed at one of the slits, the situation changes, and we now have a different point B. Point B is now at the detector, and a new path proceeds from the detector to the screen. In this eventuality there is only empty space between (B =) A' and the new terminus B', no double slit in the way, and so an interference pattern no longer appears.

When Observed Emission by Emission

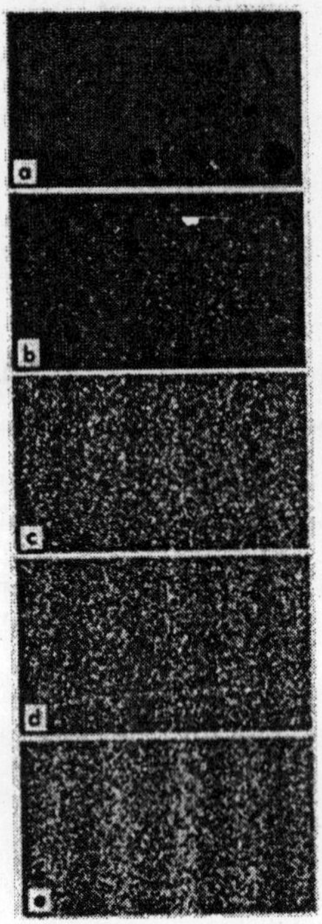

Electron buildup over time

Regardless of whether it is an electron, a proton, or something else existing on what is considered a "quantum" scale, *where* it will arrive at the screen is highly determinate (in that quantum mechanics predicts accurately the probability that it will arrive at any point on the screen). However, *in what sequence* members of a series of singly emitted things (e.g., electrons) will arrive is completely unpredictable. The experimental facts are so highly reproducible that there is virtually no argument about them, but the appearance of there being an uncaused event (because of the unpredictability of the sequencing) has aroused a great deal of cognitive dissonance and attempts to account for the sequencing by reference to supposed "additional variables".

For example, when electrons are fired at the target screen in bursts, it is easy to account for the interference pattern that results by assuming that electrons that travel in pairs are interfering with each other because they arrive at the screen at the same time, but when laboratory apparatus was developed that could reliably fire single electrons at the screen, the emergence of an interference pattern suggested that each electron was interfering with itself; and, therefore, in some sense the electron had to be going through both slits.

For something that most people continue to imagine to be an unimaginably small particle to be able to interfere with itself would suggest that this "sub-atomic particle" was in two places at once, but that idea is strongly at odds with the truism, "You cannot be in two places at once," (see law of noncontradiction). It was easier to conceptualize the electron as a wave than to accept another, more disturbing implication (from the point-of-view of our everyday notions of reality): that quantum objects are able to exist and behave in ways that defy classical interpretation.

When the double-slit experiment is performed one electron at a time with sensitive apparatus the same interference pattern emerges that would be seen if multiple electrons were fired simultaneously as had always been done with the cruder previously available apparatus. So the appearance of an orderly and consistent universe was maintained, albeit one in which everything with atomic dimensions had to be conceived as having some sort of wave nature.

However, when one electron (proton, photon, or whatever) is fired at a time, it also becomes possible to detect the point on the screen at which it arrives—and another result was demonstrated that could not easily be squared with experience of the macro world, the world of everyday experience.

In everyday experience we are accustomed to a seemingly analogous result. If one tests a firearm by locking it in a gun mount and firing several rounds at a target, a scatter pattern of bullet holes will appear in the target. We know from long experience that a poorly made gun firing poorly made ammunition will scatter shots fairly widely. We can learn and understand how flight path deviations are caused; more exacting construction of both firearms and ammunition leads to tighter and tighter patterns of bullet holes. But that is not what happens in the new double-slit experiment.

Returning again to electrons, when electrons are fired one at a time through a double-slit apparatus they do *not* cluster around a single point diametrically opposite the emitter but instead one by one fill in the same old interference pattern with which we have now become quite familiar. However, they do not arrive at the screen in any predictable order. In other words, knowing where the last electron appeared on the screen tells us nothing about where the next electron will hit.

The electrons (and the same applies to photons and to anything of atomic dimensions used) arrive at the screen in an unpredictable and arguably *causeless* random sequence, and the appearance of a causeless selection event in a highly orderly and predictable formulation of the by now familiar interference pattern has caused many people to try to find additional determinants in the system which, were they to become known, would account for why each impact with the target appears.

ELITZUR-VAIDMAN BOMB-TESTING PROBLEM

In physics, the Elitzur-Vaidman bomb-testing problem is a thought experiment in quantum mechanics, first proposed by Avshalom Elitzur and Lev Vaidman in 1993. It employs a Mach-Zehnder interferometer for ascertaining whether a measurement has taken place.

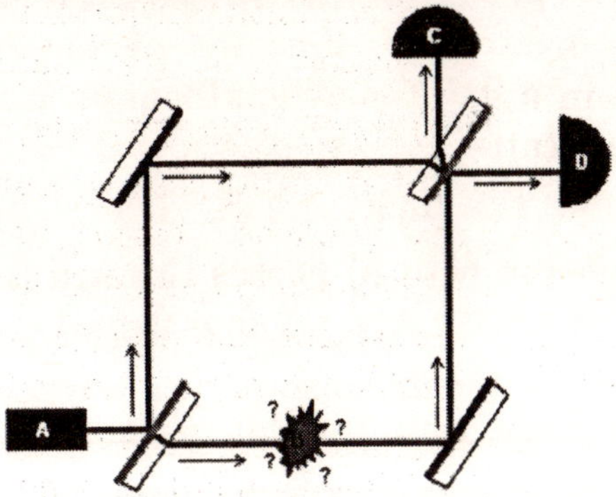

Bomb-testing problem diagram

The following example illustrates the bomb-testing problem: Consider a collection of bombs, some of which are duds. The bombs are triggered by a single photon. Usable bombs will absorb the photon and detonate. Dud bombs will not absorb the photon. The problem is how to separate the usable bombs from the duds. A bomb sorter could accumulate dud bombs by attempting to detonate each one. Unfortunately, this process destroys all the usable bombs.

A solution is for the sorter to use a mode of observation known as counterfactual measurement, which relies on principles of quantum mechanics.

Start with a Mach-Zehnder interferometer and a light source which emits single photons. When a photon emitted by the light source reaches a half-silvered plane mirror, it has equal chances of passing through or reflecting. On one path, place a bomb (B) for the photon to encounter. If the bomb is working, then the photon is absorbed and triggers the bomb. If the bomb is non-functional, the photon will pass through the dud bomb unaffected.

Due to the effects of quantum superposition, the interference effects of all possible paths taken are still noticed by the photon unless an observer collapses the wave function, at which time the photon must either exclusively be on the observed path, or exclusively be on a different path.

If the bomb is a dud:

- The photon both (i) passes through the 1st half-silvered mirror and (ii) is reflected.
- The bomb will not absorb a photon, and so the lower route is a possible path to the point of interference.
- The system reduces to the basic Mach-Zehnder apparatus with no sample, in which constructive interference occurs along the path horizontally exiting towards (D) and destructive interference occurs along the path vertically exiting towards (C).
- Therefore, the detector at (D) will detect a photon, and the detector at (C) will not.

If the bomb is usable:

- The photon both (i) passes through the 1st half-silvered mirror and (ii) is reflected.
- Upon meeting the observer (the bomb), the wave function collapses and the photon must be either on the lower route or on the upper route, but not both.
- If the photon actually takes the lower route:
    - o Because the bomb is usable, this photon triggers the bomb and it explodes.
- If the photon actually takes the upper route:
    - o Since the lower route is not taken, there will be no interference effect at the 2nd half-silvered mirror.

- o The photon on the upper route now both (i) passes through the 2nd half-silvered mirror and (ii) is reflected.
- o Upon meeting further observers (detector C and D), the wave function collapses again and the photon must be either at detector C or at detector D, but not both.

Therefore, there are only three observable results:

1. The bomb explodes.
2. The bomb does not explode and only detector (C) detects the photon. The bomb must be usable.
3. The bomb does not explode and only detector (D) detects the photon. It is possible that the bomb is usable or that it is a dud.

In the case of the third observation, the experiment may be repeated to see if the bomb will explode or if detector (C) will detect a photon. On average, this will identify all of the dud bombs, explode half of the usable bombs, and identify half of the usable bombs without detonating them.

In 1994, Anton Zeilinger, Paul Kwiat, Harald Weinfurter, and Thomas Herzog actually performed an equivalent of the above experiment, proving interaction-free measurements are indeed possible. In 1996, Kwiat et al. have devised a method, using a sequence of polarising devices, that efficiently increases the yield rate to a level arbitrarily close to one.

PHOTON DYNAMICS IN THE DOUBLE-SLIT EXPERIMENT

The dynamics of photons in the double-slit experiment describes the relationship between classical electromagnetic waves and photons, the quantum counterpart of classical electromagnetic waves, in the context of the double-slit experiment. The dynamics of a photon can be completely described by the classical Maxwell's equations with only a reinterpretation of the classical field as a probability amplitude for the photon.

Quantum Description of the Double-slit Experiment

The treatment to this point has been classical. It is a testament, however, to the generality of Maxwell's equations for electrodynamics that the treatment can be made quantum mechanical with only a reinterpretation of classical quantities. The reinperpretation is that the state vectors in the classical description of the double-slit experiment become quantum state vectors in the description of photons.

Energy and Momentum of Photons

The reinterpretation is based on the experiments of Max Planck and the interpretation of those experiments by Albert Einstein.

The important conclusion from these early experiments is that electromagnetic radiation is composed of irreducible packets of energy, known as photons.

Probability Amplitudes

The probability for a photon to be in a particular polarization state depends on the fields as calculated by the classical Maxwell's equations. The state of the photon is proportional to the field. The probability itself is quadratic in the fields and consequently is also quadratic in the quantum state of the photon. In quantum mechanics, therefore, the state or probability amplitude contains the basic probability information. In general, the rules for combining probability amplitudes look very much like the classical rules for composition of probabilities:

1. The probability amplitude for two successive probabilities is the product of amplitudes for the individual possibilities. ...
2. The amplitude for a process that can take place in place in one of several indistinguishable ways is the sum of amplitudes for each of the individual ways. ...
3. The total probability for the process to occur is the absolute value squared of the total amplitude calculated by 1 and 2.

4

Theoretical and Experimental Justification

The theoretical and experimental justification for the Schrödinger equation motivates the discovery of the Schrödinger equation, the equation that describes the dynamics of nonrelativistic particles. The motivation uses photons, which are relativistic particles with dynamics determined by Maxwell's equations, as an analogue for all types of particles.

CLASSICAL ELECTROMAGNETIC WAVES

Nature of the Photon

The quantum particle of light is called a photon. Photons have both a wave-like and a particle-like nature. In other words, photons can appear to be particles in some experiments and waves in other experiments. The dynamics of photons are completely determined by Maxwell's equations, the classical description of electrodynamics. In the absence of sources, Maxwell's equations can be written as wave equations in the electric and magnetic field vectors.

Maxwell's equations thus describe the wave-like properties of the photon. When the fields are interpreted as being proportional to probability amplitudes for finding a photon particle in a particular state, Maxwells equations also describe the particle-like properties of the photon.

Maxwell's equations were completely known by the latter part of the nineteenth century. The dynamical equations for the quantum photon were, therefore, well-known long before the discovery of the photon. This is not true for other particles such as the electron. It was surmised from the interaction of light with atoms that electrons also had both a particle-like and a wave-like nature. Newtonian mechanics, a description of the particle-like behavior of macroscopic objects, failed to describe very small objects such as electrons. Abductive reasoning was performed to obtain the dynamics of massive (particles with mass) objects such as electrons. The electromagnetic wave equation, the equation that described the dynamics of the photon, were used as a prototype for discovering the Schrödinger equation, the equation that describes the wave-like and particle-like dynamics of nonrelativistic massive particles.

PROBABILITY FOR A SINGLE PHOTON

There are two ways in which probability can be applied to the behavior of photons; probability can be used to calculate the probable number of photons in a particular state, or probability can be used to calculate the likelihood of a single photon to be in a particular state. The former interpretation violates energy conservation. The latter interpretation is the viable, if nonintuitive, option. Dirac explains this in the context of the double-slit experiment:

> *"Some time before the discovery of quantum mechanics people realized that the connection between light waves and photons must be of a statistical character. What they did not clearly realize, however, was that the wave function gives information about the probability of one photon being in a*

particular place and not the probable number of photons in that place. The importance of the distinction can be made clear in the following way. Suppose we have a beam of light consisting of a large number of photons split up into two components of equal intensity. On the assumption that the beam is connected with the probable number of photons in it, we should have half the total number going into each component. If the two components are now made to interfere, we should require a photon in one component to be able to interfere with one in the other. Sometimes these two photons would have to annihilate one another and other times they would have to produce four photons. This would contradict the conservation of energy. The new theory, which connects the wave function with probabilities for one photon gets over the difficulty by making each photon go partly into each of the two components. Each photon then interferes only with itself. Interference between two different photons never occurs."—Paul Dirac, The Principles of Quantum Mechanics, Fourth Edition, Chapter 1

Probability Amplitudes

The probability for a photon to be in a particular polarization state depends on the fields as calculated by the classical Maxwell's equations. The state of the photon is proportional to the field. The probability itself is quadratic in the fields and consequently is also quadratic in the quantum state of the photon. In quantum mechanics, therefore, the state or probability amplitude contains the basic probability information. In general, the rules for combining probability amplitudes look very much like the classical rules for composition of probabilities:

1. The probability amplitude for two successive probabilities is the product of amplitudes for the individual possibilities. ...
2. The amplitude for a process that can take place in place in one of several indistinguishable ways is the sum of amplitudes for each of the individual ways. ...
3. The total probability for the process to occur is the absolute value squared of the total amplitude calculated by 1 and 2.

de Broglie Waves

In 1923 Louis de Broglie addressed the question of whether all particles can have both a wave and a particle nature similar to the photon. Photons differ from many other particles in that they are massless and travel at the speed of light. Specifically de Broglie asked the question of whether a particle that has both a wave and a particle associated with it is consistent with Einstein's two great 1905 contributions, the special theory of relativity and the quantization of energy and momentum. The answer turned out to be positive. The wave and particle nature of electrons was experimentally observed in 1927, two years after the discovery of the Schrödinger equation.

BOHR ATOM

Inconsistency of Observation with Classical Physics

The de Broglie hypothesis helped resolve outstanding issues in atomic physics. Classical physics was unable to explain the observed behaviour of electrons in atoms. Specifically, accelerating electrons emit electromagnetic radiation according to the Larmor formula. Electrons orbiting a nucleus should lose energy to radiation and eventually spiral into the nucleus. This is not observed.

Atoms are stable on timescales much longer than predicted by the classical Larmor formula.

Also, it was noted that excited atoms emit radiation with discrete frequencies. Einstein used this fact to interpret discrete energy packets of light as, in fact, real particles. If these real particles are emitted from atoms in discrete energy packets, however, must the emitters, the electrons, also change energy in discrete energy packets? There is nothing in Newtonian mechanics that explains this.

The de Broglie hypothesis helped explain these phenomena by noting that the only allowed states for an electron orbiting an atom are those that allow for standing waves associated with each electron.

PATH INTEGRAL FORMULATION

The path integral formulation of quantum mechanics is a description of quantum theory which generalizes the action principle of classical mechanics. It replaces the classical notion of a single, unique history for a system with a sum, or functional integral, over an infinity of possible histories to compute a quantum amplitude.

The path integral formulation was developed in 1948 by Richard Feynman. Some preliminaries were worked out earlier, in the course of his doctoral thesis work with John Archibald Wheeler.

This formulation has proved crucial to the subsequent development of theoretical physics, since it provided the basis for the grand synthesis of the 1970s called the renormalization group which unified quantum field theory with statistical mechanics. If we realize that the Schrödinger equation is essentially a diffusion equation with an imaginary diffusion constant, then the path integral is a method for the enumeration of random walks.

For this reason path integrals had also been used in the study of Brownian motion and diffusion before they were introduced in quantum mechanics.

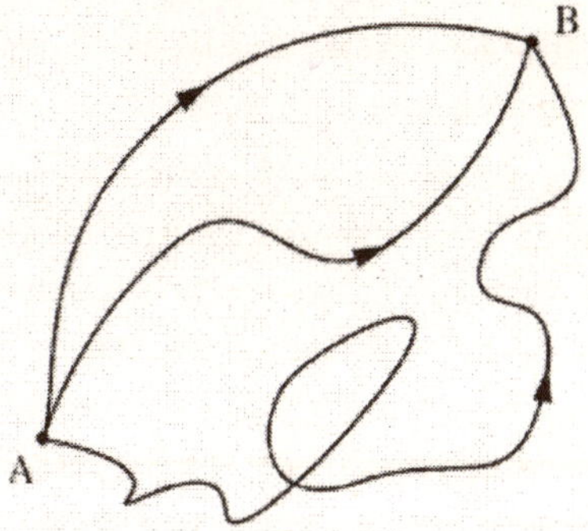

These are just three of the paths that contribute to the quantum amplitude for a particle moving from point A at some time t_0 to point B at some other time t_1.

FORMULATING QUANTUM MECHANICS

The path integral method is an alternative formulation of quantum mechanics. The canonical approach, pioneered by Erwin Schrödinger, Werner Heisenberg and Paul Dirac paid great attention to wave-particle duality and the resulting uncertainty principle by replacing Poisson brackets of classical mechanics by commutators between operators in quantum mechanics. The Hilbert space of quantum states and the superposition law of quantum amplitudes follows. The path integral starts from the superposition law, and exploits wave-particle duality to build a generating function for quantum amplitudes.

Abstract Formulation

Feynman proposed the following postulates:

1. The probability for any fundamental event is given by the square modulus of a complex amplitude.

2. The amplitude for some event is given by adding together all the histories which include that event.
3. The amplitude a certain history contributes is proportional to , where is reduced Planck's constant and S is the action of that history, or time integral of the Lagrangian.

In order to find the overall probability amplitude for a given process, then, one adds up, or integrates, the amplitude of postulate 3 over the space of *all* possible histories of the system in between the initial and final states, including histories that are absurd by classical standards. In calculating the amplitude for a single particle to go from one place to another in a given time, it would be correct to include histories in which the particle describes elaborate curlicues, histories in which the particle shoots off into outer space and flies back again, and so forth. The path integral includes them all. Not only that, it assigns all of them, no matter how bizarre, amplitudes of *equal magnitude*; only the phase, or argument of the complex number, varies. The contributions wildly different from the classical history are suppressed only by the interference of similar histories (see below).

It should be noted however, that the mathematical technique of path integrals, does not imply that real particles must actually follow the paths so constructed. Mathematical expansions of functions by other functions are a general technique, and as such the functions used are not required to have any physical interpretation at all. They are usually constructed for mathematical convenience, with no necessary analogy to the physical model that they are modeling. Indeed, in the case of the Feynman path integral, the integration is over imaginary time, so the relevance of the paths to the particle's real physical path is open to debate.

Feynman showed that his formulation of quantum mechanics is equivalent to the canonical approach to quantum mechanics. An amplitude computed according to Feynman's principles will also obey the Schrödinger equation for the Hamiltonian corresponding to the given action.

Recovering the Action Principle

Feynman was initially attempting to make sense of a brief remark by Paul Dirac about the quantum equivalent of the action principle in classical mechanics. In the limit of action that is large compared to Planck's constant , the path integral is dominated by solutions which are stationary points of the action, since there the amplitudes of similar histories will tend to constructively interfere with one another.

Conversely, for paths that are far from being stationary points of the action, the complex phase of the amplitude calculated according to postulate 3 will vary rapidly for similar paths, and amplitudes will tend to cancel. Therefore the important parts of the integral—the significant possibilities—in the limit of large action simply consist of solutions of the Euler-Lagrange equation, and classical mechanics is correctly recovered.

Action principles can seem puzzling to the student of physics because of their seemingly teleological quality: instead of predicting the future from initial conditions, one starts with a combination of initial conditions and final conditions and then finds the path in between, as if the system somehow knows where it's going to go. The path integral is one way of understanding why this works.

The system doesn't have to know in advance where it's going; the path integral simply calculates the *probability amplitude* for a given process, and the stationary points of the action mark neighborhoods of the space of histories for which quantum-mechanical interference will yield large probabilities.

CONCRETE FORMULATION

Feynman's postulates are somewhat ambiguous in that they do not define what an "event" is or the exact proportionality constant in postulate 3. The proportionality problem can be solved by simply normalizing the path integral by dividing the amplitude by the square root of the total probability for something to happen (resulting in that the total probability given by all the normalized amplitudes will be 1, as we would expect). Generally speaking one can simply define the "events" in an operational sense for any given experiment.

The equal magnitude of all amplitudes in the path integral tends to make it difficult to define it such that it converges and is mathematically tractable. For purposes of actual evaluation of quantities using path-integral methods, it is common to give the action an imaginary part in order to damp the wilder contributions to the integral, then take the limit of a real action at the end of the calculation. In quantum field theory this takes the form of Wick rotation.

There is some difficulty in defining a measure over the space of paths. In particular, the measure is concentrated on "fractal-like" distributional paths.

The Path Integral in Quantum-mechanical Interpretation

In one philosophical interpretation of quantum mechanics, the "sum over histories" interpretation, the path integral is taken to be fundamental and reality is viewed as a single indistinguishable "class" of paths which all share the same events. For this interpretation, it is crucial to understand what exactly an event is. The sum over histories method gives identical results to canonical quantum mechanics, and Sinha and Sorkin (see the reference below) claim the interpretation explains the Einstein-Podolsky-Rosen paradox without resorting to nonlocality.

Some advocates of interpretations of quantum mechanics emphasizing decoherence have attempted to make more rigorous the notion of extracting a classical-like "coarse-grained" history from the space of all possible histories.

5

NanoTechnology and the Fight Against Terrorism

NANOTECHNOLOGY

NanoTechnology involves development of materials (and even complete systems) at the atomic, molecular, or macromolecular levels in the dimension range of approximately 1-500 nanometers.Current research looks to provide detailed understanding of unique properties that materials exhibit at the NanoScale.Current focus of the research is positioned to create and use structures, devices, and systems that have unique and often contradictory properties as well as enhanced functions because of their small and/or intermediate size.NanoTechnology research and development includes control at the NanoScale and integration of NanoScale structures into larger material components, systems, and architectures as well as automated systems for the production of NanoMaterials and the automatic assembly of structures and systems.

HISTORY OF NANOTECHNOLOGY

NanoTechnology first gained recognition after Nobel Laureate, Richard Feynman, presented his talk, "There's Plenty of Room at the Bottom" to the American Physical Society in 1959.Activity surrounding NanoTechnology began to slowly increase over the next few decades.In 1988, Eric Drexler taught the first course on NanoTechnology.In that program, he suggested the possibility of nanosized objects that were self-replicating. Slow growth in this area of scientific investigation continued.The next major milestone was when Rice University Professor Richard Smalley won the 1996 Nobel Prize for discovering a new form of carbon: a molecule of sixty carbon atoms (referred to as C60).Today C60 has become one of a growing number of building blocks for a new class of nanosized materials.

The advancements in NanoTechnology really began to accelerate in the late 1990s.NanoTechnology captured the thoughts and imagination of scientists and venture capitalists after an entire issue of Scientific American addressed the promise of this new technology.

CURRENT STATE OF NANOTECHNOLOGY

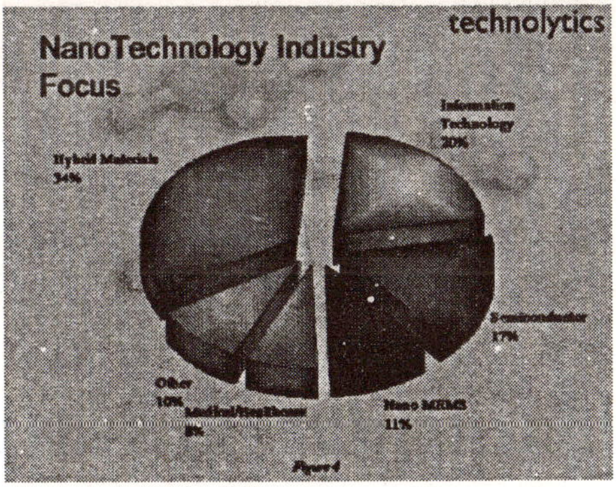

Research into NanoMaterials spans a significant spectrum of areas. Advanced material companies are producing innovative products in areas such as coatings, industrial powders, chemicals, and carbon nanotubes. Today, real world application of NanoTechnology exists in commercial business. About two-dozen serious applications of NanoMaterials and process have been fielded ranging from non-scuff floor tile to high strength brackets for running boards on vehicles to high temperature protective materials for spacecraft.While NanoMaterials are a significant portion of today's focus, several other areas are equally as promising.

TRENDS AND IMPACT OF NANOTECHNOLOGY

NanoTechnology has the potential to become a more significant revolutionary force for business than the industrial revolution or the information technology revolution.In fact, many believe that the combined impact of both the industrial and information revolution may approach the magnitude of change that could result from the commercialization of NanoTechnology.Currently, NanoTechnology is moving from the basic research stage of its evolution into the applied research stage of technology maturity.Today there are several NanoTechnology companies already being traded on the public marketplace. As this technology evolves and matures, you can expect to see many more companies enter this space.

Today's manufacturing methods are very crude at the molecular level. Casting, grinding, milling, and even lithography move atoms in mass.It is like trying to make things out of LEGO blocks with boxing gloves on your hands.Yes, you can push the LEGO blocks into great heaps and pile them up; but you cannot really snap them together the way you would like.

Future Advances

- Self-assembling products, systems and materials (self repairing system)
- Computers billions of times faster (reaching the size barrier of microprocessors)
- Extremely novel inventions (self-adapting fabrics)

- Realistic space travel and exploration (more economic)
- NanoMedical (unique treatment delivery capabilities)
- Molecular food syntheses (end of famine and starvation)

The big question is "When will this technology enter into the mainstream?" As the graphic below indicates, it will take another eight to ten years. But, rest assured, several breakthroughs and advances will produce significant advantages in materials science, semiconductors, optical communications, and information technology.

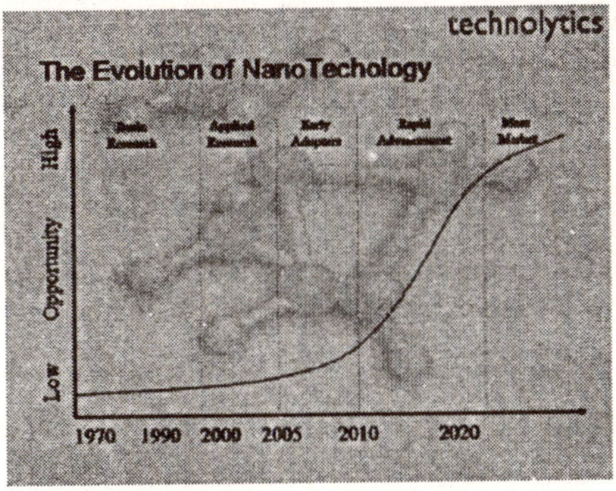

HOMELAND SECURITY APPLICATIONS

With the broad reach NanoTechnology has in terms of capabilities, the direct applications for Defense and Homeland Security are only limited by our imagination and how rapidly the technology advances.From shape-shifting armor to fabric that can turn away microbes as well as bullets to new power sources, the defense industries are launching major initiatives and planning for NanoTechnology.The Government is the major source of funding for current NanoTechnology initiatives.Centers of Excellence in NanoTechnology have been established around the country.The basic research in NanoTechnology conducted at these centers will provide the foundation upon which real world applications can be built.Other centers are already concentrating on military application of NanoTechnology.While there are efforts for new and improved weapons based on NanoTechnology, the vast majority of the NanoTechnology research and applied research fall into the support category.

From improved powers sources and batteries to advanced arming fuses the defense industry and homeland security has great interest in how NanoTechnology their capabilities.They believe that NanoTechnology will advance sensors and protective equipment and will greatly assist them in their mission. In fact, current research and development efforts are working on micro power generating devices that can be used in a plethora of applications. If successful these MPG can provide enough energy to power sensors and sensor networks that will be deployed to protect critical infrastructure like water treatment plans, roadways, and bridges.Other applications could include such things as microscopic self-powered reconnaissance and surveillance devices like listening devices, vibrations sensors as well as supplying power to sensor networks.Currently, DoD is funding research in small-scale energetic device development and management.

The lifecycle development and support requirements for NanoMEMS (micro-electro-mechanical systems) in support of advanced applications in munitions and armament systems are critical to advances in weaponry.NanoMEMS has direct implications on integrated circuits, optical switches, pressure sensors, as well as the processes used to produce weapons.However, current thoughts are that NanoMEMS can play a major role in new weapons and intelligence and surveillance. Imagine using NanoMEMS to construct a steam powered electrical generator systems for satellites.Steam is currently used in conventional and nuclear power plants to turn turbines that generate electricity.

When used onboard a satellite, as the satellite spins and out of direct sunlight, the steam condenses to liquid; and when it rotates back in the sunlight, the liquid turns to steam again.The micro-size steam power plant would generate enough electricity to operate the unit.The satellite's overall demand for energy would have been greatly reduced because of the use of extremely efficient electronics created with NanoTechnology.In addition, the overall size of the units could be reduced by a factor of ten or more, as well as the weight because of construction practices at the NanoScale.Just think! A satellite so small that it could be launched by what appears to be a large gun in place of conventional rockets.The cost savings would be enormous as well as the difficulty in anyone being able to detect the satellite in orbit.

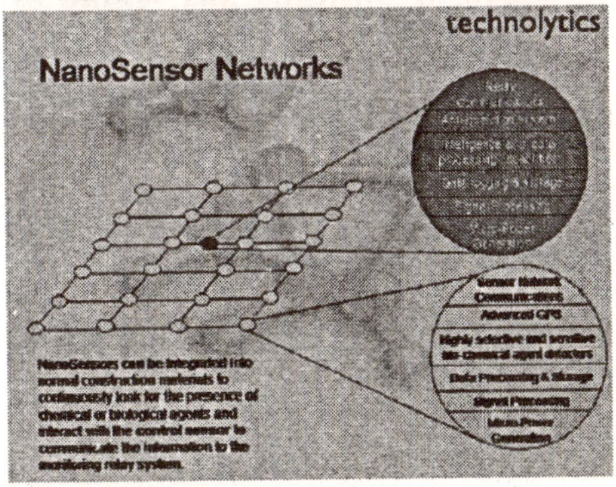

As mentioned previously, the area that NanoTechnology that holds the most significant promise is in the area of sensors and sensor networks. Hybrid NanoMaterials will produce orders of magnitude improvements in high-selectivity and high-sensitivity sensors for biological and chemical detection. This advanced detection of harmful chemical and biological agents; microsensors for radioactivity, low-power or self-powering consumption security electronics, polymer electronics, nano-optics will provide capabilities that are not available today.

Just imagine. In a few years advanced sensor networks self powered with the smarts to communicate from sensor to sensor and have the ability to detect very small amounts of chemicals or biological agents installed in the water supplies across the country.Once a single sensor node detects the presence of one of the agents, it communicates to the others what hit was found and receives their verification.Once verified, the information is communicated to the control sensor that relays the information back to the National Infrastructure Protection Center for immediate action. Other applications like underwater sensor networks to detect the movement of ships into and out of our ports could also be advanced using NanoTechnology. Sensor networks that detect chemical, biological or radiological materials could be built into cargo containers.

The applications of NanoMaterials go far beyond semiconductors and sensors into NanoFabrics. NanoFabrics with unique properties are under development at this time.Properties like decreased receptivity to chemical or biological agents, materials with the ability to expand and contract (like a thermostat) as to exhaust or conserve body heat, or to resist the penetration of a bullet.

Consider that today a soldier going into battle carries about 60 lbs of equipment.A significant part of that weight is in the bullet -proof-vest.What if the weight burden was reduced by 50%? How much more efficient would the soldier be? What if instead of having multiple types of camouflage there was only one uniform that adapted the outer colors to blend into the surroundings?

How much physical stress would be relieved if the uniform had that thermostatic characteristic discussed earlier? When the soldier becomes over heated the vents open to allow the flow of air and when the soldier becomes cold the vents close to conserve body heat.In one discussion I had preparing for this article there was even the notion that the materials could sense an injury and automatically constrict like a tourniquet or mast trousers used to treat victims in shock.

One of the early NanoTechnology successes was material that had the characteristics of Gortex with the look and feel of regular wool. Remember of course that Gortex is currently used in bulletproof vests. Today, ballistic resistant materials are heavy and many are extremely brittle. Super strength nanofabrics sandwiched into normal contraction materials are expected to improve blast resistant construction practices making commercial and governmental structures much more resilient against bomb blasts. Bomb resistant containers for cargo and luggage on ships and aircraft, bomb resistant glass for office buildings and government complexes, advanced structural members that have the strength by are pliable to absorb the energy of a blast are all currently being investigated as real world application of NanoTechnology.

Another area of NanoTechnology that is receiving the greatest amount of attention is semiconductors.The ability to further compact the number of transistors in a given space increases the performance of a semiconductor. NanoScale construction practices applied to semiconductors will substantially increase current processing capability and could change the entire industry overnight.NanoTechnology also has significant benefits in opto-electronics and communications.The ability to construct an optical switch on a chip would eliminate a significant amount of complexity and cost of optical networks, not to mention increasing the capacity as well.

CONCLUSION

While NanoTechnology is based in the research labs today, the advances made to date have illustrated the significant value this technology will bring.The implication on manufacturing techniques, materials manufacturing and the semiconductor industry will be profound.The enhanced characteristics of materials will allow us to create new and innovative devices to protect all of us from terrorism, both directly through safer construction designs and indirectly through intelligence and surveillance.With each passing day, the promise of NanoTechnology becomes increasingly apparent. There is a long way to go before the NanoMaterials production techniques can provide adequate supply of high quality materials at affordable prices. This is one technology that requires close monitoring to properly time the significant opportunities that will be created in it wake.

6

Conformal Coatings for LIGA MEMS: Issues and Challenges

MOTIVATION

Many microelectromechanical systems (MEMS) fabricated by LIGA utilize electrodeposited metals such as nickel and Ni alloys. While Ni alloys may meet the structural requirements for MEMS, their tribological (friction and wear) behavior remains somewhat undefined. For instance, the friction coefficient of pure Ni against steel and ceramic counterfaces under dry sliding conditions (m = 0.6 -1.2) is much higher than the DP design engineers would prefer.

Additionally, generation of wear debris and the stick-slip behavior could interfere with the performance and reliability.

In a number of microsystems applications such as gear trains, comb drives and transmission linkages, tribological considerations, particularly sliding contacts amongst sidewalls, is of paramount importance.

With the increasing number of MEMS fabricated using LIGA and other thick resist electroforming processes, there exists a need for the development of suitable coatings to combat the tribological problems.

ACCOMPLISHMENT

There are a variety of techniques (e.g., burnishing, resin bonding, physical vapor deposition, chemical vapor deposition) to deposit friction-reducing coatings on conventional machine elements. However, the miniature nature of LIGA MEMS elements—several hundred microns to millimeters in size—poses a tough challenge to the coating technology. Handling of individual LIGA MEMS elements is a nightmare, and applying such techniques as resin bonding and burnishing can be ruled out. Physical vapor deposition is typically a line-of-sight process, whereas the coatings are most needed on the sidewalls. Chemical vapor deposition, though conformal in nature, is a high temperature process that could potentially alter the microstructure of the base Ni alloy.

So, we have devised a novel strategy to overcome these difficulties. Instead of handling the individual LIGA MEMS elements, we coated the entire wafer before releasing the parts, i.e. after electroforming, lapping and dissolving the PMMA mold material. We opted for a commercially available plasma enhanced CVD process, in which ionized species are deposited onto biased substrates, thereby limiting the substrate temperatures to below 200°C. As a first step in this direction, we obtained a 100 nm thick commercial PECVD diamond like nanocomposite (DLN) coating on a LIGA wafer with MEMS elements on it.

The coated wafer was backsputtered to remove the coating material from the planar surfaces of the MEMS elements and from the rest of the wafer. The parts were subsequently released per standard LIGA process. TEM of a FIB section taken from a location similar to the one marked (a LIGA gear sidewall) showed that backsputtering did not impact the coating integrity on the sidewall and EDS analysis revealed all the peaks from the DLN coating. Parallel tribological measurements of DLN coatings on test coupons showed much improved tribological behavior with friction coefficient of 0.04 and practically no signs of debris generation or stick-slip behavior.

CONFORMAL COATINGS FOR LIGA MEMS: ISSUES AND CHALLENGES

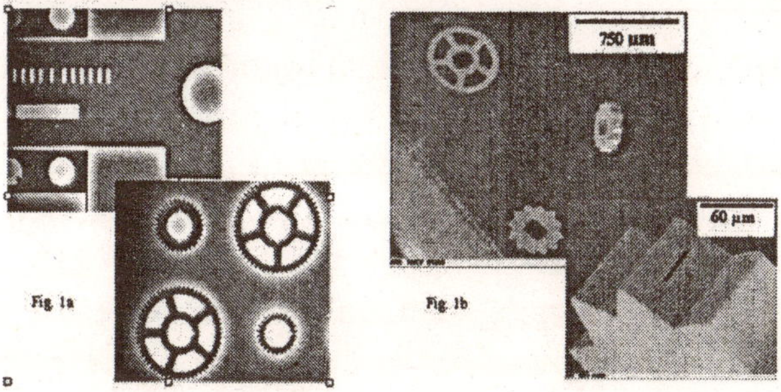

(a) LIGA fabricated components unreleased from the wafer that were coated with a 100 nm thick Diamond Like Nanocomposite (DLN) film by plasma enhanced CVD process. Figure 1b shows the SEM images of the released components after backsputtering. A FIB section was taken in a location similar to the one marked in the bottom right micrograph of the gear tooth for TEM analysis.

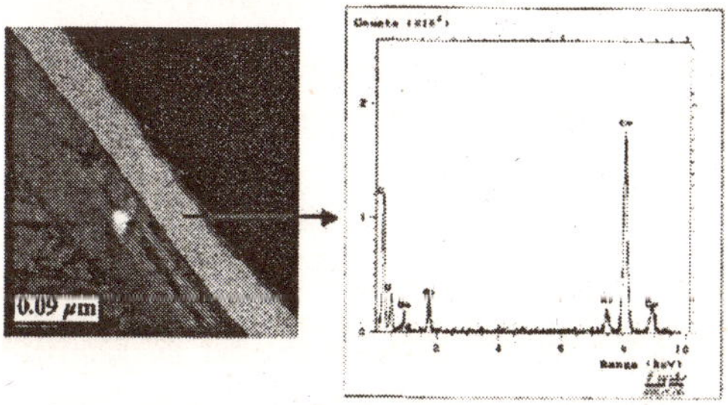

Typical TEM image of a FIB cross section from the sidewall of a DLN coated LIGA gear. The EDS spectrum shown in the right has the peaks (O, C, and Si)

corresponding to the three elements in DLN. The coating is approximately of 100 nm thick.

ANTI-STICTION COATINGS IN MEMS DEVICES

There have been many exciting predictions that the future of micromachines or microelectromechanical systems (MEMS) is just "around the corner", but this future has proven to be slow in coming. Despite the demonstration of numerous MEMS devices and product concepts each year, a very small number have actually succeeded in the market place. Two prominent examples are the digital mirror device (DMD) of Texas Instruments, Inc. and the MEMS accelerometer of Analog Devices, Inc. While these two products are vastly different, one can identify a common ingredient in their formula of success: both companies successfully developed special surface coating technologies. Indeed, the difficulty in controlling surface forces is a critical impediment to the fabrication and operation of many MEMS devices. This is a consequence of the scaling law: the surface-to-volume ratio scales with the inverse of device dimension and surface forces dominate at length scales < 1mm.

A well-known problem in the fabrication of MEMS devices from surface micromachining is stiction, which occurs when surface adhesion forces are higher than the mechanical restoring force of the micro-structure. When a device is removed from the aqueous solution after wet etching of an underlying sacrificial layer, the liquid meniscus formed on hydrophilic surfaces pulls the microstructure towards the substrate and stiction occurs. While this release-stiction problem may be alleviated by dry HF etching or supercritical CO_2 drying, a more difficult problem is in-use stiction which occurs during operation when microstructures come into contact (intentionally or accidentally).

In-use stiction may be caused by capillary forces, electrostatic attraction, and direct chemical bonding. To circumvent the stiction problem, many MEMS developers are forced to switch to bulk micro-machining, which is less capable and versatile than surface micromachining in terms of device function. Even for devices from bulk-micromachining, in-use stiction is still of concern, perhaps to a lesser extent.

One attractive approach to tackle the stiction problem is to provide low-energy surface coating in the form of an organic passivation layer on the inorganic surface. Such a coating can not only eliminate or reduce capillary forces and direct chemical bonding, but also reduce electrostatic forces if the thin organic layer is directly applied to the semiconducting substrate, without the intervening oxide layer. Texas Instruments uses a fluorinated fatty acid self-assembled monolayer (SAM) on the aluminum oxide surface in their DMD, while Analog Devices coats the surfaces of their inertia sensors using thermal evaporation of silicone polymeric materials at the packaging stage after the device is completely released. Another much advocated approach is the formation of siloxane self-assembled monolayers (SAMs) on the oxide terminated surface, but the difficulty of this chemistry and the poor reproducibility put significant limitations on its practical usage.

In view of the critical importance of anti-stiction coatings in MEMS products, we are developing a number of chemical processes that possess the following attributes:

(1) the chemistry is simple and reproducible;

(2) the coatings are of monolayer nature and are covalently bonded to the substrate;

(3) the coating processes are compatible with dry or aqueous etching processes;

(4) the monolayers are chemically and mechanically stable under conditions of processing and operation.

The design principle of the coating process is illustrated in scheme 1. A key component in the coating chemical is a molecule containing two major parts, *R & X*. The R group is selected to provide low surface energy, i.e., "wax" or "Teflon" like, while the X group is chosen to selectively react with the solid surface of interest for covalent linkage.

The attachment of these molecules to the solid surface is a specially designed process which provides kinetic control; the reaction self-terminates after a saturated monolayer coverage is reached.

This is very different from self-assembled monolayers (SAMs). Such a selective and kinetically controlled reaction ensures that the coating is uniform and conformal with solid surfaces in a MEMS device. The thickness of the coating is chosen to be 1-2 nm.

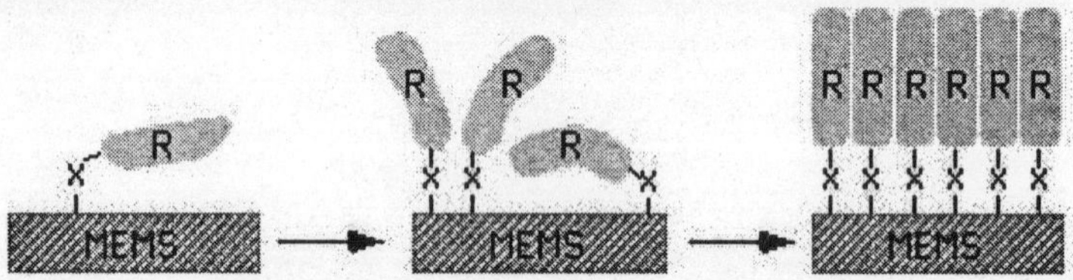

Scheme 1. Design of low energy surface coating for MEMS devices. R represents a molecular group to give low surface energy and X is a functional group for selective attachment to the solid surface.

Figure 1 shows a group of poly-Si cantilever beams completed released by the above coating process. The cantilever beam arrays (CBAs) are fabricated by the SUMMiT process of Sandia National Laboratory. A cross section of the structure is shown schematically in the upper part of the figure. The poly-Si beams, with thickness of 2.25 mm, are anchored at one end. The sacrificial PSG layer under each beam is ~2 mm thick.

The bending of p-Si beams on each CBA sample is characterized by interference microscopy. Out-of-plane bending of the beams is shown as optical fringes; each fringe corresponds to the bending of l/2 or 311 nm. The lower part of figure 1 shows typical images of beam arrays (2 mm in length) with alkoxyl coating. All beams are completely released.

The fringes reflect (downward) bending due to residual stress. The maximum bending is ~1.8 mm. Therefore, the above monolayer coating process eliminates release stiction. Electro-static actuation of these coated beams also reveals little in-use stiction with relative humidity as high as 90%.

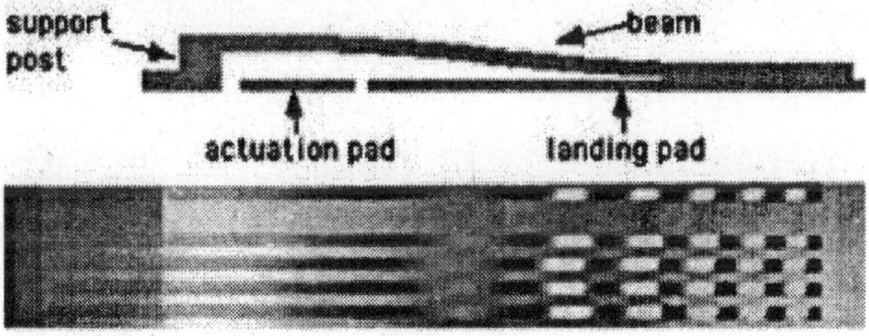

Fig. 1. Upper: schematic illustrate of a cantilever; lower: Interference microscope image of cantilever beam arrays with beam length of 2 mm. The image (composite of three frames) is obtained after the beams are released in an all-liquid etching-coating process.

NEW BIOMECHANICAL PHENOMENON USING TINY SILICON "FINGERS"

"Micromachines" could Lead to New Medical Treatments, Nano-robots

As reported in the April 14 issue of the journal *Science*, the researchers discovered that DNA bends tiny silicon "fingers" that have a thickness of less than a 1/50 of a human hair. Using a process called "molecular recognition," in which molecules bind according to a key-lock mechanism, an array of fingers or cantilevers arranged like the teeth of a comb was made attractive to specific DNA codes and proteins.

By observing the way different cantilevers bent as the DNA adhered to them, the researchers were able to detect the tiniest possible defect in a DNA sequence, a so-called single base mismatch.

Advantages ot the new technology are direct detection of the targeted substance, and it has good sensitivity even for small molecules.

"This biomechanical technique has the potential to enable fast and cheap biochemical analysis, and could be used for mobile applications," said Christoph Gerber of IBM Research. "For example, there is a need for rapid diagnosis of the condition that is a leading cause of sudden death — the heart attack.

If the technology can be developed as we hope, it could make it possible to determine on the spot if chest pain is caused by a heart attack or a more benign problem, saving time and potentially lowering treatment costs substantially."

The team's discovery that DNA and proteins will bend tiny silicon structures such as cantilevers has much wider implications, and could make it possible to extend the technology to the operation of micro- and nano-robotic machinery.

"Microbots and nanobots have been popularized in recent science fiction stories and movies, but technological issues remain an obstacle to their realization," said James Gimzewski of IBM Research. "The ability to use biology to perform specific mechanical tasks on the nanometer scale with silicon provides a completely new approach to operate machinery autonomously, without external power or computer control. We have found a way to get DNA to do the work for us, so we don't need batteries, motors, or the like to operate tiny machines."

The ability of DNA and other biomolecules to operate machinery such as valves using the molecules' specific code or biochemistry could have applications in medicine. "For instance, we can envision a system to attack cancerous growth: the release of just the proper doses of chemicals in the appropriate location of the body could be achieved using tiny microcapsules equipped with nano-valves," said Gimzewski.

"They could be programmed chemically to open only when they get biochemical signals from a targeted tumor type. This would enable the right therapy at the right place at the right time, with minimized side effects and no invasive surgery."

The technology underlying this new advance springs from development work on nanomechanical olfactory sensors, as pursued by the research teams at IBM in Zurich and at the University of Basel*. To date, applications of this work are mainly in quality and process control, where the technology is used in sensing devices for gaseous analytes, such as process gases or solvent vapors.

Such devices are not limited to gaseous environment, but also function in liquids. This led the way to the research now reported in *Science* on using the biomechanical sensors in biochemistry and medical diagnostics.

TECHNICAL BACKGROUND

The method is based on directly transforming specific biochemical recognition into a nanomechanical motion.

To do this, the researchers used hybridization, the base pairing between two single strands of DNA that results in the well-known double-helix structure. Hybridization is a prominent example of molecular recognition.

The core of the device is an array of silicon cantilevers, each 500 microns long, 100 microns wide and less than 1 micron thick. Each cantilever is coated on one side by specific biomolecules.

When immersed in solution, molecules of an injected substance dock to a layer of receptor molecules that have been attached to one side of the cantilever.

Sensitizing an array of cantilevers, each with a different receptor, allows docking of different substances in the same solution. The increase of the molecular 'packing density' leads to surface stress and thus to bending of the cantilever.

The bending is of the order of 10-20 nanometers, which can be measured accurately by well-established methods, such as laser beam deflection. The hybridization was done with short strands of single-stranded DNA (12mer oligonucleotides) and proteins known to recognize antibodies of various mammals.

The scientific report on this work is published in *Science*, Vol. 288, Number 5464, p. 316ff., April 14, 2000. The authors of the report "Translating Biomolecular Recognition into Nanomechanics" are Juergen Fritz, Marko Baller, Hans Peter Lang, Ernst Meyer and Hans-Joachim Guentherodt of the University of Basel, and Hugo Rothuizen, Peter Vettiger, Christoph Gerber, and James Gimzewski of IBM's Zurich Research Laboratory.

The work was partially funded by research programs of the Swiss Government (MINAST -Micro and Nanosystems Technology- and TOPNANO21 **) for micro- and nanosystems.

ILLUSTRATIONS

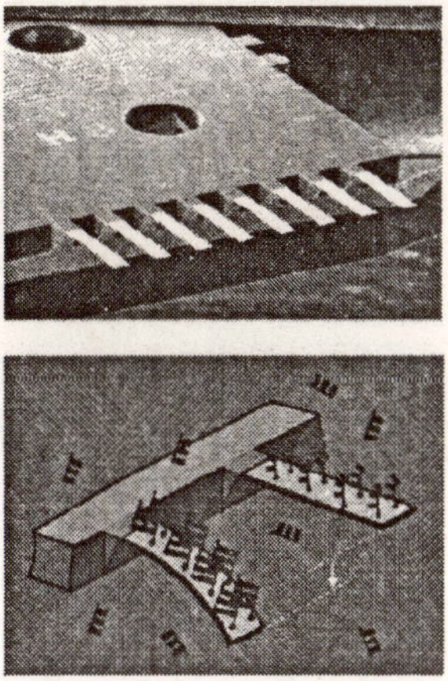

Figure 1) shows an array of tiny cantilevers imaged by scanning electron microscopy. The cantilevers bend by docking of molecules (see schematic).

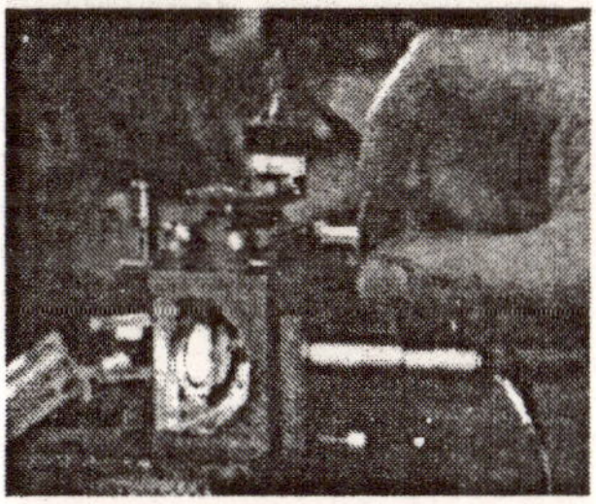

Figure 2) shows the experimental setup of the biosensor in the laboratory.

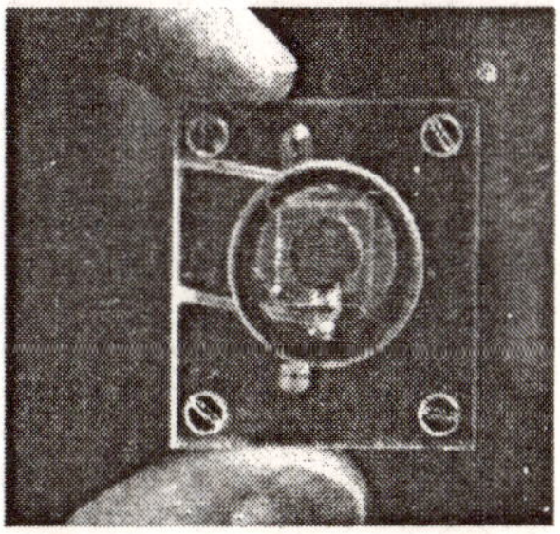

Figure 3) The core of the instrument is the liquid cell in which the cantilevers are mounted.

Additional images

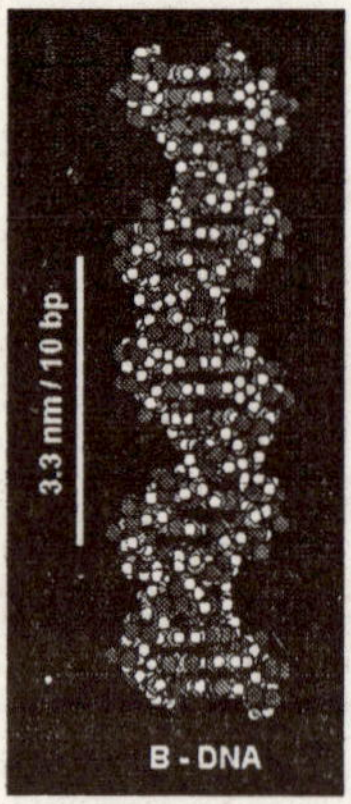

Figure 4) DNA forming the well-known double helix structure.

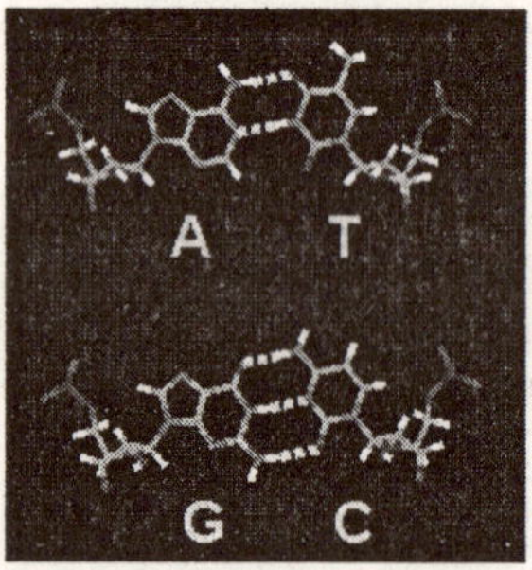

Figure 5) The code letters of DNA - the bases A, T, G and C - forms well defined bonds. A always binds to T, G always binds to C. This process is called hybridization.

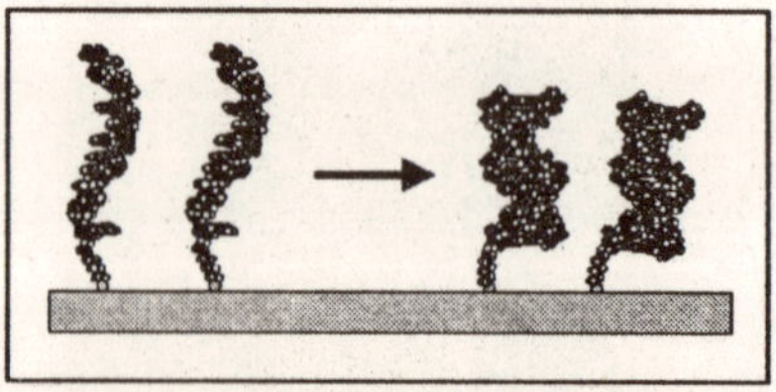

Figure 6) Single strands of DNA recombine via hybridization to double stranded DNA (double helix structure). The authors demonstrated that hybridization of DNA on a gold coated cantilever produces bending of the cantilever.

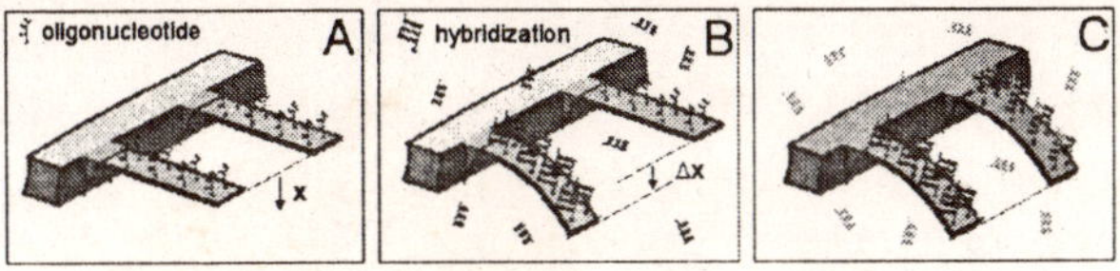

Figure 7) Schematic drawing of the experiment. A. Two cantilevers covered by single stranded DNA. B. Bending of the first cantilever by hybridization. C. Bending of the second cantilever by hybridization.

MONOLAYER PROTECTIVE COATINGS

Self-healing bound and mobile phase lubricants extend operational life of microelectromechanical systems.

Scientists at the Materials and Manufacturing Directorate, working with the University of Dayton Research Institute (UDRI), developed bound and mobile phase monolayer protective coatings capable of extending the operational life of microelectromechanical systems (MEMS) used on aerospace and space systems. The directorate's Nonmetallic Materials Division developed and tested the new coatings with technical assistance from UDRI as part of a research effort to determine the durability and failure mechanisms of MEMS operating in extreme environments (i.e., vacuum, high temperature, and moist air).

The conclusions drawn from the research, thus far, demonstrate that monolayer lubricants with bound and mobile phases provide self-healing surfaces and are highly effective in reducing friction, wear, stiction, adhesion, and other tribological phenomena hindering the development and sustained operation of MEMS devices on current and future Air Force systems.

MEMS, an enabling technology for a number of systems important to the Air Force and national security, are particularly crucial for miniaturization and integration, and enhancing capabilities in spaceborne military systems. They are also very important to commercial industry. Examples include miniature satellites, actuators, gyroscopes, microwave switches, accelerometers, sensors, airflow control, aircraft turbine engines, and unmanned air vehicles. MEMS are extremely small, batch-fabricated systems comprised of electrical and mechanical components on the millimeter to nanometer scale.

Their principal advantage over conventional systems is an ability to significantly reduce the size, weight, power consumption, and cost of aerospace and space systems. MEMS technology and processes include lithography, bulk micro machining, surface micro machining, etching, film deposition, and packaging. Examples of structural materials used to build them include polysilicon, silicon carbide, silicon nitride, and diamond.

MEMS can enhance capabilities affecting flight safety and mission success. This is particularly true for onboard diagnostics and reconfiguration of microsatellites. Researchers understand that protecting them against friction, wear, stiction, adhesion, and other phenomena that hinder their performance and shorten operating life is a formidable challenge. Directorate scientists and engineers sought to overcome this challenge and established a research program to understand and modify lubricant structure at the molecular level. Their subsequent research effort, called "Tribology in Extreme Environments," focused on three primary concerns—novel thin film deposition, friction and wear fundamentals, and new materials development—and soon led to major reductions in lubricant sensitivity to humidity.

Following up on their initial success, the research team directed its efforts toward optimizing bound and mobile phase lubricants capable of providing self-healing surfaces for MEMS. Since this lubricant system is very effective at elevated temperatures, in the vacuum of space, and in high humidity environments, it offered real potential for optimizing MEMS performance, benefitting the Air Force.

Working with UDRI, directorate researchers successfully demonstrated Fomblin® Z-DOL 2000, a lubricant system with bound and mobile phases, on a miniature electrostatic lateral output motor (see figure). They chose the motor as the test base because of its large electrode area, which provides a sufficient level of force, and because it has a number of different contact interfaces for tribological studies and ensures the relevancy of any lubrication scheme. The output motor also provides a platform for several interesting experiments. They chose Fomblin Z-DOL 2000 because it provides the proper bound and mobile phase mix after being heat-treated, and it has already proven successful as a computer hard disc lubricant. Bound monolayer lubricants wear away quickly. The research team demonstrated that the addition of a mobile phase creates a lubricant system with a self-healing capability. This new system may hold the key to building MEMS that perform better and last longer.

MEMS electrostatic lateral output motor used to test bound and mobile phase lubricants

7

Advances in Transportation Due to Nanotechnology

Nanotechnology will impact in all areas of transportation, providing a variety of tools to build the system for the twenty-first century. For example, advances in nanotechnology may make possible carbon-based fibres which are 100x stronger than steel, at only one-sixth the weight.

THE BENEFITS OF ADVANCED MATERIALS

Advanced materials that will allow for longer service life and lower failure rates. Among the key applications are nanocoating of metallic surfaces to achieve super-hardening, low friction, and enhanced corrosion protection; 'tailored' materials for infrastructure and vehicles; and "smart" materials that monitor and assess their own status and repair any defects resulting from fatigue, fire, etc. New materials designed to have superior properties (stronger, lighter) using materials with nanoscale dimensions will lead to faster, cheaper, and safer transportation

The Benefits of Advanced Communications

Advanced communications that maximise the benefits of intelligent transportation systems and obviate the need for some travel altogether; sensors that continually monitor the condition and performance of roads, bridges, and other infrastructure This includes:

- New materials that will permit ultra-miniaturisation of space systems and equipment, including the development of 'smart' sensors and probes. Applications include economical supersonic aircraft; low-power, radiation-hardened computing systems for autonomous space vehicles; and advanced aircraft avionics
- Reduction in transportation energy use and its impacts on the environment. Applications include nanosensors to monitor vehicle emissions and trigger traps for pollutants; nanoparticle-reinforced materials that replace metallic components in cars; replacement of carbon black in tyres with nanoparticles of inorganic clays and polymers, leading to tyres that are environmentally friendly and wear-resistant; and carbon-based nanostructures that serve as 'hydrogen supersponges' in vehicle fuel cells.

IMPROVEMENT IN PUBLIC SAFETY

Such applications may include smart sensors for collision avoidance in automobiles; nano-reinforced, glare-resistant car windows and windshields; chemical and biological weapon detectors at airports; and sensors that monitor rail track structural integrity This includes:

- Improved catalysts could reduce or eliminate the emission of pollutants from engines
- New photonic nanodevices could replace the heavy and costly radio frequency transmission equipment on board aircraft, ships, or satellites or be exploited for inexpensive remote control of vehicles

NANOTECHNOLOGY UNDERPINNING DEVELOPMENTS IN THE AUTOMOTIVE INDUSTRY

Nanotechnology is underpinning developments in the automotive industry in a number of areas. The automotive industry is a major user of sensors and components for new integrated miniaturised systems, and MEMS have been a key driver in many of today's advanced safety systems. MEMS are currently used in two standard automotive applications: the air bag accelerometer, which first went small tech in the early 1990's, and the manifold air pressure sensor, or MAP sensor, first used in the late 1970's. Total automotive sales of silicon accelerometers are expected to increase from £192m in 2000 to £364m in 2005 (www.strategyanalytics.com). The worldwide MST market (including MEMS) for automotive applications in 2000 was £725m and is projected to grow to £800m by 2004. (NEXUS/Roger Grace Associates).

PHOTONICS

Photonics is the science of generating, controlling, and detecting photons, particularly in the visible and near infra-red spectrum, but also extending to the ultraviolet (0.2 - 0.35 μm wavelength), long-wave infrared (8 - 12 μm wavelength), and far-infrared/THz portion of the spectrum (e.g., 2-4 THz corresponding to 75-150 μm wavelength) where today quantum cascade lasers are being actively developed. Photonics is an outgrowth of the first practical semiconductor light emitters invented in the early 1960s at General Electric, MIT Lincoln Laboratory, IBM, and RCA and made practical by Zhores Alferov and Dmitri Z. Garbuzov and collaborators working at the Ioffe Physico-Technical Institute and almost simultaneously by Izuo Hayashi and Mort Panish working at Bell Telephone Laboratories. Photonics most typically operates at frequencies on the order of hundreds of terahertz.

Just as applications of electronics have expanded dramatically since the first transistor was invented in 1948, the unique applications of photonics continue to emerge. Those which are established as economically important applications for semiconductor photonic devices include optical data recording, fiber optic telecommunications, laser printing (based on xerography), displays, and optical pumping of high-power lasers. The potential applications of photonics are virtually unlimited and include chemical synthesis, medical diagnostics, on-chip data communication, laser defense, and fusion energy to name several interesting additional examples.

RELATIONSHIP TO OTHER FIELDS

Classical Optics

Photonics is closely related to optics. However optics preceded the discovery that light is quantized (when the photoelectric effect was explained by Albert Einstein in 1905). The tools of optics are the refracting lens, the reflecting mirror, and various optical components which were known prior to 1900. The key tenets of classical optics, such as Huygens Principle, the Maxwell Equations, and wave equations, do not depend on quantum properties of light.

Modern Optics

Photonics is approximately synonymous with quantum optics, quantum electronics, electro-optics, and optoelectronics. However each is used with slightly different connotations by scientific and government communities and in the marketplace. Quantum optics often connotes fundamental research, whereas photonics is used to connote applied research and development.

The term photonics more specifically connotes:

(1) the particle properties of light,

(2) the potential of creating signal processing device technologies using photons,

(3) those quantum optical technologies which are manufacturable and can be low-cost, and

(4) an analogy to electronics.

The term optoelectronics eponymously connotes devices or circuits comprising both electrical and optical functions, i.e., a thin-film semiconductor device. The term electro-optics came into earlier use and specifically encompasses nonlinear electrical-optical interactions applied, e.g, as bulk crystal modulators such as the Pockels Cell, but also includes advanced imaging sensors typically used for surveillance by civilian or government organizations.

Emerging Fields

Photonics also relates to the emerging science of quantum information in those cases where it employs photonic methods. Other emerging fields include opto-atomics in which devices integrate both photonic and atomic devices for applications such as precision timekeeping, navigation, and metrology. Another emerging field is polaritonics which differs with photonics in that the fundamental information carrier is a phonon-polariton, which is a mixture of photons and phonons, and operates in the range of frequencies from 300 gigahertz to approximately 10 terahertz.

Overview of Photonics Research

The science of photonics includes the emission, transmission, amplification, detection, modulation, and switching of light.

Photonic devices include optoelectronic devices such as lasers and photodetectors, as well as optical fiber, photonic crystals, planar waveguides, and other passive optical elements.

Applications of photonics include light detection, telecommunications, information processing, illumination, metrology, spectroscopy, holography, medicine (surgery, vision correction, endoscopy, health monitoring), military technology, laser material processing, visual art, biophotonics, agriculture and robotics.

History of Photonics

Photonics as a field really began in 1960, with the invention of the laser, and the laser diode followed in the 1970s by the development of optical fibers as a medium for transmitting information using light beams, and the Erbium-doped fiber amplifier. These inventions formed the basis for the telecommunications revolution of the late 20th century, and provided the infrastructure for the internet.

Historically , the term photonics only came into common use among the scientific community in the 1980s as fiber optic transmission of electronic data was adopted widely by telecommunications network operators (although it had earlier been coined). At that time, the term was adopted widely within Bell Laboratories. Its use was confirmed when the IEEE Lasers and Electro-Optics Society established an archival journal named Photonics Technology Letters at the end of the 1980s.

During the period leading up to thc dot-com crash circa 2001, photonics as a field was largely focused on telecommunications. However, photonics covers a huge range of science and technology applications, including:

- laser manufacturing,

- biological and chemical sensing,
- medical diagnostics and therapy,
- display technology,
- optical computing.

Various non-telecom photonics applications exhibit a strong growth particularly since the dot-com crash, partly because many companies have been looking for new application areas quite successfully. A huge further growth of photonics can be expected for the case that the current development of silicon photonics will be successful.

Applications of Photonics

- Consumer Equipment: Barcode scanner, printer, CD/DVD/Blu-ray devices, remote control devices
- Telecommunications: Optical fiber communications
- Medicine: correction of poor eyesight, laser surgery, surgical endoscopy, tattoo removal
- Industrial manufacturing: the use of lasers for welding, drilling, cutting, and various kinds of surface modification
- Construction: laser levelling, laser rangefinding, smart structures
- Aviation: photonic gyroscopes lacking any moving parts
- Military: IR sensors, command and control, navigation, search and rescue, mine laying and detection
- Entertainment: laser shows, beam effects, holographic art
- Information processing
- Metrology: time and frequency measurements, rangefinding

- Photonic computing: clock distribution and communication between computers, circuit boards, or within optoelectronic integrated circuits; in the future: quantum computing

A TECHNOLOGY FOR TOMORROW

The different stages of technological evolution are characterized in terms of the dominant type of tools used in each of them. We have passed trough the 'stone age', the 'copper age' and the 'iron age' and are probably now in the 'electron age'. After having done a wonderful job, today it looks as if electronic technologies have started experiencing their limits. For example, today's 'information explosion' and communication problems demand very large bandwidths without cross talk, larger than that electronics can provide. What could be the technology that can tackle the growing needs of tomorrow?

Photonics is being studied today as a possible alternate technology for the future. Investigations are on to harness the capability of the photon to carry information and energy. Fortunately, success has already been achieved in the area of communications. Optical fibers, not copper cables, are already being used to carry huge amounts of information across the great oceans. Even the internet cables around us are now optical and not electronic.

However, communication is only one of the three 'C' s, that has been taken care of, the other two being computing and control. Photonic switches and optical computers are still in the laboratories and not yet in the marketplace. The major bottleneck in this area is not in solving technological problems or in perfecting theoretical understanding, but in developing suitable materials. That is where contributions from Materials Scientists are called for.

The Nature of Optical Nonlinearity

Optical processes of materials have always attracted the attention of Materials Scientists. Spectroscopic characterization and analysis of materials using the techniques of optical absorption, luminescence, Raman scattering are standard techniques in research. These studies help in structure analysis and in understanding the electronic processes and energy levels in systems. Several interesting photochemical processes and reactions such as photosynthesis have attracted the attentions of Materials Scientists and Biologists. The advent of lasers has revolutionized optical technology including spectroscopic instrumentation.

The high intensity radiation from lasers is also capable of causing new processes to occur in materials. In such cases, most of the materials can have a 'nonlinear interaction' with the electric field. The nonlinear interaction results in several novel processes, which have the potential for communication, control and computing applications.

Essentially, the proportionality of the induced electrical polarization P in the medium to the electric field E breaks down and the resulting polarization can be considered to be made up of several terms consisting of products of higher order susceptibility c(n) and the magnitude of the electric field E. This mathematical formalism helps us to classify optical nonlinearity of materials and to present several important aspects of it in a convenient way.

P and E are vector quantities and c(n) is a tensor of rank (n+1), with 3(n+1) individual components. These tensor components describe the directional dependence of optical properties of crystals. The second and subsequent terms inside the brackets for the expression for susceptibility are progressively much smaller than the first term.

This means that nonlinear optical effects would vanish in the low optical intensity regime as intensity is proportional to the square of the amplitude E of the electromagnetic wave. A material can be expected to exhibit n th order optical nonlinearity when either of the quantities c(n) or E is large enough. E depends on the intensity of the laser used and c(n) is a property of the material. Thus the amount of nonlinearity induced will depend on both the nature of the material as well as the intensity of the laser used.

From Optics to Photonics

The nth term in the above expression for c(n) describes an '(n+1) wave-mixing' phenomenon. According to this formalism, the first order nonlinearity (n=1) actually corresponds to linear optics and hence c(1) governs most of the ordinary optical phenomena such as reflection, refraction, diffraction, interference, polarization etc. In linear optics, we have the principle superposition of waves according to which light waves passing through a medium do not exchange energy with one another. This is not the case for n>1, and for n=2, we have wave mixing phenomena such as second harmonic generation (SHG), second order sum-and-difference frequency generation etc. SHG refers to the generation of light at frequency 2w when light of frequency w interacts with a medium.

For example, infrared light of wavelength 1060 nm from an Nd:Yag laser could be converted to the green light of wavelength 532 nm. In the case of frequency mixing, light at frequencies w1 and w2 can generate light at w1 ýÿw2 in a second order medium, in addition to light at 2w1 and 2w2. A smart combination of such processes in a proper medium can ultimately provide several laser wavelengths by process known as optical parametric mixing. This is the technology being widely used by the latest models of high power tunable lasers in the market today.

Second order phenomena are NOT exhibited by materials which possessing a center of inversion. The existence of inversion symmetry forbids c(2) -related processes in materials.

This restricts second order materials to certain classes of crystals. This is not the case with third order phenomena, which could be exhibited by all materials. Apart from higher order wave mixing phenomena, several new processes are also possible here. The refractive index of the material n becomes a function of the intensity of light I according to the relation n(I) = n0 + n2 (I) where n0 is the linear refractive index and n2I is the nonlinear contribution. This in fact makes the medium act as a lens when a strong beam of light with a Gaussian cross section passes through it.

This is because the central part of the beam sees a larger effective refractive index and consequently travels slower, when compared to the peripheral part. Some materials self-focus the beam whereas some others self-defocus, depending on whether n2 is positive or negative. The change in refractive index with intensity is the basic principle based on which several photonic devices such as optical switches, transistors, modulators couplers, limiters etc.

Several methods have been designed to measure the nonlinear optical susceptibilities in materials. These are based on the optical processes involved. For example, third order nonlinear susceptibility can be measured conveniently using the self-focusing effect. In an experiment known as z-scan, a laser beam with Gaussian cross section is passed through a convex lens and the far field beam pattern is studied as the medium is moved along the beam direction, across the focal point of the lens. Obviously, the beam profile will change if the medium also starts acting as another lens. Thus a plot of the intensity (at any point) in the far field cross section of the beam as a function of the position of the sample enables us to calculate the beam distortion and hence the third order nonlinear susceptibility.

The dreams of Photonics Technology would be realized only if proper materials with large nonlinear susceptibility and fast response apart from ease of preparation and handling could be developed. Most of the natural materials such as inorganic crystals, organic liquids, semiconductors etc. have only either one of these conditions satisfied and not both simultaneously. This is a frontier area of intense research in the area of developing new Photonic Materials and improving the nonlinear optical response of known materials.

8

Quantum Confinement Effects in Semiconductor Nanostructures

Spatial confinement of electrons in semiconductor nanostructures leads to remarkable changes in their quantum states. This results in considerable modifications in their electronic properties and consequently their linear and nonlinear optical response, thus making them very attractive candidate materials from the points of view of both fundamental Physics as well as their Technological potential.

Quantum wells, which are two-dimensional confined systems, have been studied and utilized for practical applications to a large extent these days whereas not that much attention has been being paid to the interesting properties of quantum wires and dots. Quantum dots are quasi zero-dimensional systems which are practical realizations of the familiar text book concept of particle-in-a-box. One way of realizing this is by preparing semiconductor islands of nanometer dimension in dielectric matrices such as glasses and polymers. The dielectric provides the confining potential and also acts as a convenient embedding matrix for applications. Nanostructures of several materials such as PbS, CdS and CdSe have been prepared in glass and polymer matrices in this manner and several interesting results have been obtained. As an example, some recent results on CdS quantum dots in Nafion polymer are presented.

PREPARATION AND CHARACTERIZATION

CdS quantumm dots are prepared in Nafion polymer by an ion-exchange reaction. Starting with a cadmium salt solution, the undoped polymer is kept immersed in it to facilitate the exchange of cadmium ions into Nafion. The initial concentration of cadmium ions in the solution and the duration of treatment determine the mean size of the nanoclusters thus formed. The polymer sheets are then exposed to hydrogen sulphide gas for appropriate duration, when CdS forms in the form of nanoclusters in the naturual pores of the polymer structure. The size can be determined by X-ray Diffraction and electron microscopy.

SPECTROSCOPIC STUDY OF BANDGAP ENGINEERING

Quantum confinement effects become pronounced in the region of strong confinement, where the size of the quantum dot is less than the exciton Bohr diameter of the semiconductor. This parameter has a value of 6 nm for CdS, The band picture is no longer strictly valid in the case of quantum confinement and transitions between excitonic levels gain in oscillator strength even at room temperature. These effects can be easily seen in optical absorption and photoluminescence spectra.

The band edge exhibits a signature blue shift with decreasing dot size. Also, excitonic transitions show up as multiple peak structures in optical absorption and luminescence excitation spectra. this is 6 nm. The emission spectrum shows the characteristic exciton peak even at room temperature. Since optical densities are quite large, the excitonic features are often masked by the strong absorption in the optical absorption spectra. However, photoacoustic spectroscopy can help to reveal the features of excitonic transitions and to probe the energy level structure.

POTENTIAL AS PHOTONIC MATERIALS

One of the main motivations of studying nanostructures is their potential as materials for Photonic applications. Frontier research in Photonics revolves around development and characterization of materials with large and fast nonlinear optical susceptibilites. Quantum dot structures have been known as potential candidate materials in this regard though much has not been done about systematic characterization of the size dependence of nonlinearity especially in the strongly confined regime. Optical phase conjugation studies on CdS quantum dots reveal strong enhancement in third order susceptibilities with decreasing dot size. Moreover, it appears that the nonlinearity is of predominantly non-thermal origin, which is an attractive feature for potential applications.

PROBLEMS AND PROSPECTS

Perspective of Photonic Materials

Progress in Photonics over the last decade has indeed been fascinating. However, there exists a serious limitation in exploiting the potential of Photonic processes to their fullest extent. This is known as the 'materials bottleneck . It turns out that most of the available materials do not meet the simultaneous requirements of large as well as fast nonlinearity required for the practical devices. Several inorganic materials such as KDP, ADP, barium titanate etc, have been used from the earliest days. Organic materials and semiconductors also have been found to be excellent materials. Special molecules such as fullerenes and porphyrins are also studied as promising materials.

Examples of Molecular Engineering

Some specific examples here may help to illustrate the potential of an active role for of the Materials scientist in present day scenario of Photonics. The first case is that of molecular engineering in organic polymers. The structures of certain polymers such as polyphenyl acetylene (PPA) have well-spread electron clouds resulting in large values of linear as well as nonlinear susceptibility. Structure of such polymers can be modified easily by doping with other materials and/or attaching appropriate side groups to the main chain. Some of these variations are known to enhance the nonlinearity of the basic polymer by way of extending the spatial spread of electrons and hence enlarging the linear as well as nonlinear susceptibilities.

A second example is that of structure variation studies in metalloporphyrins which also has yielded interesting results by way of enhancement of optical nonlinearity and low laser power applications. Porphyrins allow for incorporation of various ligands and different core metal ions, which alter the susceptibilities dramatically. Incorporation of porphyrins into solid polymer membranes is found to enhance nonlinearity, besides stabilizing the materials against photodegradation and providing a convenient solid matrix.

Examples of Band Gap Engineering

One of the frontier area of Photonic materials development today is that of 'Quantum Engineering' of semiconductor nanostructures. Quantum wells are made of thin alternate layers of semiconductors of different band gaps, thus confining the electron to virtually a plane. These are now being used in several devices. Electrons can be further confined to one dimension by embedding a semiconductor wire or a conducting polymer chain in a material of larger band gap. Research on quantum wires and nanotubes is being pursued actively these days.

Quantum dots can be formed by embedding tiny crystalline islands of semiconductors or a metals (called nanoclusters) in glass or polymer matrices. These are practical realizations of 'particle in a box' whose optical properties depend on the particle size. Some of the leading techniques of obtaining such monosize clusters have been perfected by Materials Scientists. These include ion-exchange strategies and control of cluster formation in chemical reactions by capping with surfactants. Other techniques are ion-implantation, monolayer building and molecular beam epitaxy, which also are being used widely to produce strongly confined quantum dots.

Quantum confinement alters the electronic structure of the semiconductor and hence influences the linear and nonlinear optical properties. The band gap is modified by the increased oscillator strength of exciton transitions and hence there is an effective blueshift of the absorption edge with decreasing cluster size.

This results in an alteration of the effective band gap, which is also known as 'Bandgap Engineering'. Recent studies indicate that strong quantum confinement results in considerable increase in the nonlinear susceptibilities along with a decrease in response times. Collaborative attempts by Materials Scientists and physicists along these directions are sure to be productive in designing novel types of photonic materials ideally suited for exploiting the fascinating possibilities of Photonics to their fullest extent.

AN OVERVIEW OF PHOTOACOUSTIC SPECTROSCOPY

Salient Features

Photoacoustic spectroscopy (PAS) is a novel spectroscopic technique with several distinct features that make it a complementary technique to the conventional forms of spectroscopy. This is basically a branch of photothermal spectroscopy in which the heat generated due to the absorption of light energy in a sample is measured directly. In contrast, it may be noted that the optical absorption properties of a sample are studied by actually measuring the light transmitted by it in the conventional optical absorption spectroscopy. In PAS, the sample is periodically heated by the incidence of a chopped light beam and the pressure wave thus generated in the coupling gas enclosing the sample is detected by acoustic techniques. Such a scheme of experimentation provides several interesting special features, which make PAS a very attractive tool for the Chemist, the Physicist, the Materials Scientist and even the industrialist [1,2].

These features include

a) the basic simplicity of the experimental set up,

b) the convenience of studying samples in a variety of physical forms - crystals, powders, thin films, liquids etc.,

c) the possibility of investigating opaque materials,

d) the feasibility of direct measurement nonradiative lifetimes,

e) the ability to gather information from sub-surface layers,

f) the absence of stringent requirement on sample preparation and

g) the scope to obtain information on thermal parameters of the sample.

In fact this technique has been successfully used in a wide variety of applications ranging from semiconductor characterization to eye-lens degradation studies. Let us see the basic principles of this form of spectroscopy that makes all this possible.

Basic Concepts

The processes that occur in a solid on being illuminated by light depend very much on the nature of the sample and of course, the wavelength of the light incident. Absorption, transmission, reflection, scattering and fluorescence are the possible processes that can occur. Absorption could be followed by partial radiative reemission. Nonradiative decay channels also are possible, which eventually result in an increase in the temperature of the sample.

This heating up would depend on the (angular) chopping frequency w, the optical absorption coefficient b and the thermal diffusion coefficient a of the sample. Two characteristic lengths may be defined as Lop = (1/b) and Lth = (1/v{2a/w}) which typically are measures of the lengths over which the optical and thermal effects can have their influence. A theoretical analysis of PA response based on the thermal and optical properties of the sample has been given by Rosencwaig and Gresho.

In the case of optically transparent samples, Lop is greater than the sample thickness d and hence light is absorbed by the entire length of the sample. Such samples are considered to be thermally 'thin' if Lth is greater than both Lop and d. Then the PA signal intensity is directly proportional to (db/w) where w is the (angular) chopping frequency of incident light. For thermally thick samples, only the light absorbed in the first layers of the sample contribute to the signal and the signal is proportional to at the product (b Lop)w.

Thus we see that by varying the chopping frequency we can actually probe different layers of the sample, thus offering the possibility for depth-profiling. By increasing the frequency we are effectively increasing the thermal diffusion length Lth and hence observing PA signal originating from a deeper layer within the sample. PA depth profiling opens up exciting possibilities such as studying layered and amorphous materials and determination of thickness of thin films. In the case of optically opaque (Lop<d) samples, the PA signal strength varies as w-1 . Even for opaque samples, b can be evaluated by analyzing the PA signal whereas conventional optical spectroscopy is not possible.

EXPERIMENTAL SET-UP

The heart of the experimental set of a PA spectrometer is the PA cell. It encloses the sample in a leak-free gas atmosphere at a desired pressure. Mostly, air at atmospheric pressure is used due to the simplicity of experimentation. However, better signals can be obtained if the coupling has a pressure slightly above the atmospheric pressure. The sample holder is designed such that light can be incident on the sample through a window. The cell volume is kept a minimum for better signal strength. The cell is to be isolated from possible acoustic disturbances.

The incident light from a white light source such as a xenon lamp of a few hundred watts power is chopped at a desired frequency using a mechanical chopper. The absorption of chopped light by the sample causes periodic heating, which alters the size of the sample periodically, leading to alternate condensation and rarefaction in the coupling gas. This pressure wave, normally in the acoustic frequency region, can be easily detected by a powerful microphone of sensitivity of more than 10 mV/Pa. A piezoelectric detector can be glued directly onto the sample the case of solid samples or to the wall of the container in the case of liquid samples, for the purpose of better acoustic impedance matching. The actual magnitudes of the temperature-rise and size-change of the sample are very small and hence the PA signal from the acoustic detector has to be amplified. Lock-in amplification is found to be very advantageous.

For PA spectroscopy the wavelength of the incident light is varied continuously using a scanning monochromator and the signal is monitored as a function of the wavelength. This spectrum has to be corrected for the spectral distribution of the light source and the dispersion of the monochromator. Carbon black, a perfect absorber in the entire visible region, is used as a standard reference for the purpose of spectral correction. Data acquisition and processing become easier and more accurate on computer interfacing.

SCOPE AND POTENTIAL

PA response is widely used is a variety of scientific and technical applications. Spectroscopy of otherwise difficult samples (such as powders, opaque materials etc.) is an area of immediate interest to Chemists and Materials Scientists. This also provides a direct measurement for nonradiative lifetimes.

PAS has been used successfully as a spectroscopic tool for investigating and characterizing colored inorganic as well as organic materials including transition metal complexes, semiconducting materials, liquid crystals and metals. A correlative study of PAS and optical spectroscopy can be effectively used for obtaining a more complete picture of the electronic processes in materials. An area of special interest to the Chemist is the identification of species and their valence states in several situations such as catalysis and other chemical reactions. The effect of fluorescence quenchers can be studied by monitoring the PA and fluorescence signals simultaneously. Other related applications include photosynthesis and other photochemical reactions, quantum efficiency studies on organic dyes etc. The possibility of studying coloring agents such as cyctochrome and hemoglobin samples has attracted the attention of biologists. Some biomedical applications such as blood examination also have been demonstrated. Usually blood specimens have to be rendered transparent by smearing a thin layer on a glass plate after some processing in order to record the optical absorption spectrum. However, PA spectrum of whole blood as such can be recorded to get the same information as that obtainable with optical absorption. Similarly, PA spectrum of a whole leaf has been shown to provide information about coloring agents of the leaf. Another example is the study apple peal by depth profiling PA spectroscopy. Other medical applications of PAS include identification of bacterial states, study of animal and human tissues including teeth, bone, skin, muscle etc., analysis of drug in tissues, investigation of the photo-oxidative decay in human eye lenses etc.

PA imaging and microscopy have been demonstrated to be efficient tools in obtaining visual information on a microscopic scale. This is done by focusing light onto a spot on the sample and scanning it over the sample while recording the PA signal. Any local change in the thermal or optical property shows up as a change in the PA signal at that point of scanning. Depth-profiling and subsurface imaging can be done by varying the chopping frequency. This has potential as both a general analytic tool as well as a process control instrument in semicodnuctor industry.

A RECENT STUDY ON SEMICONDUCTOR NANOSTRUCTURES

Semiconductor nanostructures are of current topical interest due to the fact that their properties undergo considerable modification from those of the bulk due to quantum confinement effects. Nanoclusters of these materials embedded in dielectric hosts act are known to behave as quantum dots, described rightly as practical examples of 'particle-in-a-box'. Such structures can be prepared by several chemical and physical methods such as ion-exchange reactions, molecular beam epitaxy, ion-implantation, sol-gel techniques etc .

Quantum confinement leads to alterations in the band structure of the bulk. It also systematically shifts the band edge towards higher energy as the quantum dots get smaller and smaller. This is known as the signature 'blue shift' in optical absorption and the process as 'bandgap engineering'.

The physical processes in such structures are governed by their excitonic energy levels, which are difficult to be probed by optical spectroscopy due their proximity to band-edge absorption, as the optical densities tend to be very large and transmission poor. However, we may seek a technique that directly measures the absorbed energy to investigate these systems. PA spectroscopy has been used successfully to obtain valuable spectroscopic information in such cases.

In summary, it is quite established that the study of PA response has the potential to develop into a common useful analytic and research technique in a variety of scientific, industrial and medical applications both as a spectroscopic tool as well as a nonspectroscopic probe for thermal and elastic properties.

9

Advantages and Disadvantages of Nanotechnology

In Simple word it can be said that all technologies are double-edged swords. Fundamentally new technological capabilities and benefits are accompanied by new risks, and new responsibilities for managing risks appropriately. The acceptance of these responsibilities is not optional. Dealing with the benefits and risks of nanotechnology openly and proactively, neither amplifying nor downplaying them, will be critical to the positive development of the field.

The first version of the Foresight Guidelines was developed during and after a workshop on Molecular Nanotechnology (MNT) Research Policy Guidelines sponsored by the Foresight Institute and the Institute for Molecular Manufacturing (IMM). The workshop was conducted over the February 19-21, 1999, weekend in Monterey, California.

Participants included: James Bennett, Greg Burch, K. Eric Drexler, Neil Jacobstein, Tanya Jones, Ralph Merkle, Mark Miller, Ed Niehaus, Pat Parker, Christine Peterson, Glenn Reynolds, and Philippe Van Nedervelde. The Guidelines have been revised many times in the intervening years. The resulting Foresight Guidelines ("the Guidelines") include assumptions, principles, and some specific recommendations intended to provide a basis for the responsible development of nanotechnology.

Continued research and education are needed to create a shared understanding and sufficient knowledge base on the entire set of nanotechnology development and risk management issues that must be addressed. While discussion of guidelines can begin today, the scientific and technical community will continue to evolve its understanding of the issues. The Guidelines have already changed over time to reflect that dynamic understanding and specific feedback from a wider community.

The term nanotechnology refers to several distinct classes of technology, each with its own set of capabilities, potential applications, and risks. The specific terms used for these technologies vary over time; however, it is important to be clear about the fundamental distinctions between them.

NANOSCALE SCIENCE

The nanoscale science and engineering conducted today has been defined as technology with a size range less than 100 nanometers or billionths of a meter. Using that definition only, most of chemistry would qualify.

The practitioners in the field would add that nanoscale science and engineering researches and exploits the unique properties of materials in the less than 100 nanometer size range. For example, depending on its specific configuration, carbon in that size range may exhibit extraordinary tensile strength greater than diamond, or act as an electrical conductor, insulator, or semiconductor. This branch of nanotechnology exists today with many research programs throughout the world, and many companies commercializing their applications.

These companies may sell purified nanotubes made of tubular lattices of carbon, soccer ball like polyhedra made of 60 atoms of carbon called Buckyballs, dendritic polymers called dendrimers, or other particles in the less than 100 nanometer size range. The applications of these technologies are numerous and significant. They enable fundamentally new types of pharmaceuticals, electronic memory and semiconductor devices, sensors, renewable energy capture and storage systems, water purification devices, super strong fabrics and materials, security and military components, as well as antipollution devices. These applications are already beginning to emerge, and will gather momentum over the coming years.

The risks associated with passive compounds in the less than 100 nanometer size range concern their ability to be inhaled, absorbed through the skin, or to pass through biological compartment barriers such as the blood brain barrier. They thus pose a range of potential health and environmental risks that are associated with their potential toxicity or mutagencity in their interactions with biological systems. While the range of effects vary, most of the risks may be addressed by advanced industrial hygiene and environmental health practices and techniques that seek to characterize the specific risks, exposure patterns, and control methods and enforce them through a combination of practitioner education, industry self-regulation, monitoring and government regulation.

This is an important emerging field in the environmental and health sciences, since most of the existing legislation on environmental, safety, and health risks may cover particulates, but do not take the change in physical and biological properties at the nanoscale into account. It is reasonable to assume that passive nanoscale particle risks, although potentially serious if not addressed, will be characterized and addressed systematically under new versions or extensions to existing occupational, industrial hygiene, environmental, and medical regulations.

PRODUCTIVE NANOSYSTEMS

Productive nanosystems are currently a research oriented class of nanotechnology that will produce programmable, molecular-scale systems that make other useful nanostructured materials and devices. These systems may take over a decade to develop and mature.

They will be qualitatively different from nanomaterials, particularly regarding regulatory issues. These systems may be used as infrastructure for manufacturing, specifically the ability to build molecularly precise, inexpensive, three-dimensional products of arbitrary size.

The most straightforward infrastructure for manufacturing will be built with special purpose molecular fixtures and components that are analogous to macroscale factory components that produce devices that are inherently incapable of replication. These special purpose manufacturing systems will eventually be able to manufacture very large structures by scaling specific components and sub systems.

The simplest, most efficient, and safest approach to productive nanosystems is to make specialized nanoscale tools and put them together in factories big enough to make what is needed. People use simple tools to make more complex tools, from blacksmiths' tools to automated machinery.

The convergent assembly architecture developed by Ralph Merkle (1997 Nanotechnology 8 18-22), describes how small parts can be put together to form larger parts, starting with nanoscale blocks and progressing up the hierarchy to macroscopic systems.

The machines in this would work like the conveyor belts and assembly robots in a factory, doing similar jobs. If you pulled one of these machines out of the system, it would pose no risk, and be as inert as a light bulb pulled from its socket.

The eventual applications of these special purpose manufacturing systems include the ability to build almost any mechanical device cheaply, and in large quantity. This is why productive nanotechnology manufacturing capabilities will eventually do for our relationship to molecules and matter what the computer did for our relationship with bits and information.

The computer enabled an ever expanding number of people to access billions of dollars worth of information. Productive nanotechnology will enable an ever expanding number of people to enjoy significant material wealth, based on carbon feedstock, which currently is in overabundant supply. It will also enable the technical infrastructure to address effectively many of our most pressing transportation, environmental, medical and global warming issues.

The primary risks of manufacturing enabled by productive nanosystems concern what is manufactured, not the manufacturing infrastructure itself. Special purpose manufacturing systems can be designed to be safe and reliable.

They could be made to build a wide range of devices cheaply, in place, and on-demand. These products could include components for large scale buildings, computing, mass transit systems, energy storage, and spacecraft. On the other hand, they could also include tiny new security devices, and large quantities of inexpensive and super strong conventional weapon systems.

THE REPLICATORS

The concern about advanced forms of nanotechnology, after feasibility issues have been addressed, tends to center on the possibility of the technology getting out of control. Popular science fiction describes molecular robots that are autonomous self-replicating machines or autonomous replicators that evolve beyond human control, or can't stop reproducing. For the purpose of our Guidelines, an autonomous replicator is a specific kind of device that both (1) contains a set of materials processing and fabrication mechanisms sufficient to perform the operations necessary build devices like itself and (2) contains a set of design instructions and instruction-interpreting mechanisms sufficient to direct the operations necessary to build a device like itself. All other machines, lacking these exceptional properties, are not autonomous replicators. Productive nanotechnology enabled manufacturing of small or large products does not require any autonomous replicators, either in development or in application. It is also important to distinguish between the special purpose ability to manufacture many copies of a specific product, and the ability of the manufacturing infrastructure to replicate itself.

Nature employs its own biological form of molecular manufacturing to produce organisms. Productive nanosystems can be completely non-biological. They can be designed to function in a narrow and controlled range of physical conditions. Autonomous self-replicating assemblers are not necessary to achieve significant manufacturing capabilities. As Drexler and Phoenix indicated in their Safe Exponential Manufacturing paper (2004 Nanotechnology 15 869-872), developing manufacturing systems that use general purpose replicators able to extract their own energy sources is unnecessary.

It is important to note that molecular nanotechnology not specifically developed for manufacturing could be implemented as non autonomous replicating systems that have many layers of security controls and designed-in physical limitations. This class of system could potentially be used under controlled circumstances for nanomedicine, environmental monitoring, and specialized security applications. There are good reasons to believe that when designed and operated by responsible organizations with the appropriate quality control, these non autonomous systems could be made arbitrarily safe to operate. However, a determined and sophisticated group of terrorists or "non state entities" could potentially, with considerable difficulty, specifically engineer systems to become autonomous replicators able to proliferate in the natural environment, either as a nuisance, a specifically targeted weapon, or in the worst case, a weapon of mass destruction.

Both conventional nanoscale technology and manufacturing enabled by productive nanosystems can be implemented by responsible parties quite effectively without these risks, but as with other technologies, the risk of abuse must be considered seriously. Thus, in addition to the need for professional ethics and multiple layers of embedded industrial controls, there will also be a need for thoughtful regulation, monitoring, and potentially the development of "immune responses" to external threats.

EMBEDDED CONTROLS

There are many additional dimensions of safety controls that can be engineered into designs for replicators. Robert Freitas and Ralph Merkle (2004, Kinematic Self-Replicating Machines, Landes Bioscience, p. 152) have described a multidimensional space of 137 replicator design properties, characterized by structure, function, inputs, and outputs, and the percentage of its own components that a system can fabricate, extract, transport, inspect, warehouse, repair, control, or energize. Each of these design properties represents a potential control point for identification, embedded safety features, and deactivation. The authors support an earlier version of these Guidelines in their book, and agree with its recommendations against developing inherently unsafe replicator designs which permit: (1) surviving mutation or undergoing evolution, and (2) competing with or using biology as a raw materials resource.

Replicator designs may be made arbitrarily safe by employing layers of redundant embedded controls. For example, a broadcast architecture which transmits encrypted manufacturing instructions to a machine without an on-board instruction set. Another example is a vitamin architecture which designs in dependence on an exotic fuel or substance not available in the natural environment. While no technology is absolutely safe and free from the risk of abuse, we will refer to these systems as inherently safe replicator designs.

Specifically, they are inherently safe compared to systems that are built without redundant layers of built-in safety controls. It is important to note that molecular manufacturing systems could be made considerably safer and more vulnerable to deactivation than natural bacteria or viruses which evolve rapidly, use biological materials as food, are readily available today, and are easier to hijack for nefarious purposes. Understanding the underlying properties of different forms of non autonomous replicating systems will make it much easier to engineer them with embedded safety and security features, as well as detect and control any attempts to abuse them.

Inherently safe designs may be needed for non autonomous replicators that can monitor the environment and respond rapidly to attacks or natural disasters. The development of nanotechnology is part of a general technological trend towards miniaturization that is proceeding globally at an accelerating pace. Bad actors who might try to abuse the technology could potentially be thwarted if replicator countermeasures or an "immune system" is deployed in advance. These systems would be analogous to those developed by nature for when organisms come under attack by viruses and bacteria, and anti-virus programs developed by humans to protect computing systems. Like our own very successful but imperfect immune system, this type of system could be compromised by sophisticated technologists and programmed to attack itself, as the body does in autoimmune diseases.

This does not mean immune defenses are useless. In fact, they have been selectively developed during millions of years of evolution. It does mean that we will need to build redundant layers of security into the design of these systems, and keep improving them rapidly. This may become a "predator-prey" cycle as competitors respond to the latest defense or attack mechanisms. More research and resources are required on the large space of design for safety in non autonomous replicator systems, design for security and rapid response in immune systems, and guidelines concerning their development and use.

BENEFITS AND RISKS

Policy related discussions of nanotechnology require consideration of the economic and environmental benefits of the different types of nanotechnology, as well as the potential problems. Some of the potential benefits of advanced forms of the technology include new lifesaving systems to address massive poverty and hunger, prevent and repair damage from diseases that are not responsive to today's antibiotics or anti-viral drugs, and provide rapid material response to natural disasters such as tsunamis, hurricanes, earthquakes, and volcanoes. Poverty, disease, and natural disasters kill thousands, in some cases millions annually, and the potential to ameliorate their effects significantly should not be relinquished lightly, particularly by those least affected.

The Foresight Nanotechnology Challenges address critical needs that could be met by developing a range of near and long term nanotechnology solutions. They include:

(1) meeting global energy needs through more efficient generation, storage and distribution,

(2) providing abundant clean water through improved water purification and filtration,

(3) increasing health and longevity of human life through medical diagnostics, drug delivery and customized therapy,

(4) maximizing the productivity of agriculture through precision farming, targeted pest management and the creation of high yield crops,

(5) making powerful information technology available everywhere through reduced cost and higher performance of memory, networks, processors and components, and *6) enabling the development of space resources through improved fuels, as well as smart materials and environments.

Meeting any one of these challenges with nanotechnology would have a transformative effect on our quality of life. Thus, the need for new controls should not prevent the responsible development of the field.

There has been considerable confusion about the sources of risk in the development and future deployment of nanotechnology. Most of the risk assessment of current nanoscale technology concerns compatibility of nanomaterials with humans and the environment. Relevant ecological and public health principles must be utilized in the development of any technology, particularly one as fundamental and broad as nanotechnology. Nanoparticles may be absorbed or inhaled if the proper industrial hygiene precautions are not utilized. They may also pass readily through body compartments such as the blood brain barrier. Manufactured diamondoid products may not break down easily in the natural environment. Consumers may not initially have means readily available to recycle them. Thus, total "product lifecycle" considerations about health and environmental effects should be taken into consideration as industry develops new manufacturing techniques based on productive nanosystems.

Since some controls on the different classes of nanotechnology will eventually be put into place, it makes sense for them to be as well informed as possible. Rather than have external controls imposed upon an R&D community that is not addressing potential risks openly, the developing nanotechnology R&D community should adopt appropriate self-imposed controls proactively. They can also participate in policy discussions on external controls that may be formulated in light of current knowledge and the evolving state of the art. The quantity and quality of these additional controls will depend to some extent on the success of voluntary controls.

RISKS REDUCTION

Industry and government should have the maximum opportunity to develop and commercialize a manufacturing industry based on productive nanosystems designed for safety and reliability. In addition, nanotechnology should be developed in ways that make it possible to distribute the substantial benefits of the technology to the majority of humanity currently desperate to achieve material wealth at any environmental or security cost. Manufacturing based on productive nanosystems will eventually be capable of producing widespread material abundance with significantly less environmental impact than technologies in common use today. Providing technical abundance alone cannot make a country wealthy and secure.

This also requires education, and social arrangements that include rule of law, and other features of civil society. However, technological abundance can alleviate many of the conflicts that stem primarily from rivalry over resources. Reducing this specific cause of conflict could make the world considerably more secure than it is today. In addition, the release from bare economic subsistence could enable billions of people to take advantage of the emerging global classroom enabled by the World Wide Web. This education effect could compound the positive security benefits of nanotechnology.

Given the accelerating world wide research on various types of nanotechnology it is highly probable that advanced capabilities will eventually be developed, and thus, it is important for society to consider proactively the range of controls that should be in place to minimize risks and maximize economic benefits. Unlike the proliferation of nuclear technology, which can be partially controlled by limiting access to fissionable materials, once the technology matures, it could utilize carbon, which is readily available. Further, there is no guarantee that limiting research and development on these systems in open and democratic countries would effectively slow their development and deployment elsewhere. China, for example, has an active nanotechnology research program, and it is training large numbers of qualified scientists and engineers in the relevant disciplines for developing advanced forms of the technology. The Guidelines take the position that the safest and most responsible path is to enact reasonable technology-specific controls at the practitioner, industry, and governmental levels, and simultaneously develop the monitoring and counter measures to control potentially rogue or offensive use of the technology.

REGULATION AND TREATIES

Effective means of restricting the misuse of molecular nanotechnology in the international arena will need to be developed. If weaponized versions of MNT are developed, they may not fall under existing arms-control treaties. Adding productive nanosystems designed for manufacturing to the list of technologies covered in Chemical, Biological and Nuclear Weapons treaties could be inappropriate because MNT is not a weapon, but a productive technology with broad applications. It is more similar to chemical technology than to chemical weapons, and more similar to biotechnology than to biological weapons.

Adding particular weapons related applications of MNT to the list of technologies covered in Chemical, Biological and Nuclear Weapons treaties may be appropriate in certain cases. It should be remembered, however, that the capabilities of productive nanosystems will be extensions of general manufacturing technology. The military applications of MNT will include the manufacture of high performance aerospace vehicles and precision munitions at low cost. The high value and dual use of MNT for civilian and defense purposes will require making distinctions between the enabling technology and its specific applications, balancing health and economic benefits against security concerns. Since nanotechnology research is now global, the security challenges will be present, with or without the ability to capture the wide variety of benefits.

Overly restrictive treaties or regulatory regimes applied to core MNT technologies could lead to the unintended consequence that only the rule-following nations would be at a competitive disadvantage technologically, economically, and militarily. While most nations, companies, and individuals are likely to adhere to reasonable safety restrictions, guidelines that are viewed as too restrictive will simply be ignored, paradoxically increasing risk. In addition, not all guidelines and laws will be followed, and enforcement is rarely perfect.

Non-state actors could become quite significant, particularly when the relevant knowledge and raw materials are available globally. They may well not be signatories to any agreement. While a 100% effective ban could, in theory, avoid the potential risks of certain forms of molecular nanotechnology, a 99.99% effective ban could result in development and deployment by the 0.01% that evaded and ignored the ban. For example, the international Biological Weapons Treaty was being violated on a massive scale even before the ink was dry.

On the other hand, international cooperation on restricting the proliferation and use of atomic weapons has been partially effective in limiting their development and use over the past several decades. In that case, however, the raw materials were not widely available. There are reasonable arguments on both sides of the treaty question. It is wise to avoid an unnecessary nanotechnology arms race, particularly when manufacturing enabled by productive nanosystems could be used to greatly reduce competition for material resources and mutually improve quality of life for the competitors. Treaties may become easier to verify, at least when the verifiers are more advanced than the verified. The "trust, but verify" security concept will increase in importance in a potentially dangerous, but increasingly transparent world. In fact, nanotechnology based sensors are likely to increase multifold the options for transparency.

The Guideline participants as a group have not endorsed any specific means to address MNT security concerns through treaty arrangements. However, as nanotechnology capabilities increase, governmental, NGO, industry cooperative arrangements, and self-funding enforcement mechanisms will need to be implemented. Self policing NGOs and industry groups will play an important role in this, but the public is likely to want some oversight provided by elected representatives. This may reflect distinctions between addressing potential voluntary and localized risks vs. involuntary and more widespread risks, as well as differences in the distribution of benefits. In addition, monitoring defense applications will have to be handled primarily by government agencies. There will be no easy answers in this arena, only a set of dynamic trade-offs that will have to be incorporated into flexible organizations and policies that can survive the vagaries of the real world of politics, economics, and accelerating technology.

The international community of nations, industry, and nongovernmental organizations will need to develop effective means of enforcing nanotechnology guidelines and regulations, and responding to misuse. Such means should not restrict the development of peaceful applications of the technology or defensive measures by responsible members of the community. Today there are obvious difficulties in achieving consensus on the definition of "responsible members" and enforcing international agreements on chemical and biological weapons is known to be problematic. However, given the importance of making progress in this area, further research, collaboration, and innovation is encouraged.

ENFORCEMENT

The safe development and use of nanotechnology depends, in part, on the good judgment and ethical behavior of the researchers carrying out this work. This is an imperfect, but important first line of defense. The more this is recognized as critically important, the more effective the vast majority of researchers are likely to be in actively preventing unsafe designs or uses of nanotechnology, and in insuring that manufacturing systems have built-in safeguards.

The natural and responsible path for the development of productive nanotechnology based manufacturing makes use of no autonomous replicators. However, defense against potential rogue elements who might seek to abuse replicators is a problem not unlike the challenge of controlling the developers of viruses on the Internet. In both cases, a combination of moral and technical education, active industry and government cooperation, inherently safe system designs, legal frameworks, and R&D on secure immune systems for defense may be the best solutions available.

Nanotechnology policy will have to balance risks with benefits, and distinguish between different classes of risks. Molecular manufacturing and nanotechnology are not one technology, but rather a spectrum of technologies, with radically different risk profiles. A substantial R&D

program is needed to clarify the nature, magnitude and likelihood of the potential risks, as well as the options available for dealing with them effectively. For example, toxicology analyses relating to nanomaterials have already been identified as an early priority. Nanomaterial safety is a matter that is distinct from the key risks of productive nanotechnology based manufacturing, but both require good industrial hygiene practices.

There are significant risks associated with failing to address the increasingly costly economic, political, environmental, energy, and security problems that the development of productive nanosystems could help resolve. Likewise, there are real costs to restrictive policies that limit nanotechnology innovation by responsible actors and allow rogue entities to move ahead. The Guidelines were not intended to cover every risk or potential abuse of the technology. People may abuse automobile technology, and society has responded by making cars safer to operate, holding drivers accountable for their actions through laws that are enforced, and requiring drivers to pay for automobile insurance. Likewise, industry and governments are held responsible for their use of technologies that have widespread impact.

The Guidelines are intended to cover most of the risks associated with normal development and use of the technology, and to mitigate, as much as possible, the risks associated with potential abuse of the technology. However, most guidelines and embedded control regimes can be circumvented by a sophisticated and determined adversary. Defending against this kind of threat may require an active monitoring and immune system for detection and deactivation.

Informed policy decisions could accelerate the safe development of peaceful and environmentally responsible uses of nanotechnology. This includes capturing the opportunity to develop powerful new approaches to medicine, and energy efficient, zero emission manufacturing and transportation technologies. Informed and thoughtful policies will also increase our security by enabling new types of immune systems to detect and respond to deliberate abuse.

The field of nanotechnology is very broad, like computing, and thus its eventual regulation spans human health and safety (NIH), environmental protection (EPA), and eventually weapon systems (DoD, DHS, CIA).

It is important that appropriate distinctions be made between different classes of nanotechnology in the development of regulations, and that effective interagency coordination is done to promote and enforce consensus, and build on existing regulatory standards and monitoring where appropriate.

The self assessment scorecards are based on the notion that the people, organizations, and governments that work in the nanotechnology field should develop and utilize professional guidelines and practices.

These guidelines and practices should be grounded in science and technology principles, and knowledge of the interacting environmental, security, ethical, and economic issues relevant to the development of the field.

This is based on the notion that professional ethics, "soft law", and cultural norms regarding good practice are at least as effective as "hard law" in preventing unsafe practices, and in helping to ensure that unsafe practices are noticed and acted upon.

The use of "soft laws" is a first line of defense, and is not meant to suggest that "hard laws" for safety and health are not useful, and at times appropriate.

Any regulation adopted by researchers, industry or government should provide specific, clear guidelines. Regulators should have specific and clear mandates, providing efficient and fair methods for identifying different classes of hazards and for carrying out inspection and enforcement.

There is great value in seeking the least-restrictive necessary legal environment to ensure the safe and secure development of each specific type of nanotechnology. It is important to recognize in this context that some types of nanotechnology will eventually provide the best solutions available for remedying the existing environmental and public health damage resulting from our current, distinctly suboptimal, technology base.

10

Applications of Nanotechnology

Over the past few decades, the fields of science and engineering have been seeking to develop new and improved upon types of energy technologies that have the capability of improving life all over the world. In order to make the next leap forward from the current generation of technology, scientists and engineers have been developing a new field of science called Nanotechnology. Nanotechnology is simply any technology that contains components smaller than 100 nanometers. For scale, a single virus particle is about 100 nanometers in width.

An important subfield of Nanotechnology related to energy is Nanofabrication. Nanofabrication is the process of designing and creating devices on the nanoscale. Creating devices smaller than 100 nanometers opens many doors for the development of new ways to capture, store, and transfer energy.

The inherent level of control that Nanofabrication could give scientists and engineers would be critical in providing the capability of solving many of the problems that the world is facing today related to the current generation of energy technologies.

People in the fields of science and engineering have already begun developing ways of utilizing Nanotechnology for the development of consumer products.

Benefits already observed from the design of these products are an increased efficiency of lighting and heating, increased electrical storage capacity, and a decrease in the amount of pollution from the use of energy. Benefits such as these make the investment of capital in the research and development of Nanotechnology a top priority.

CONSUMER PRODUCTS

Recently, previously established and entirely new companies such as BetaBatt, Inc. and Oxane Materials are focusing on Nanomaterials as a way to develop and improve upon older methods for the capture, transfer, and storage of energy for the development of consumer products.

ConsERV, a product developed by the Dais Analytic Corporation, uses nanoscale polymer membranes to increase the efficiency of heating and cooling systems and has already proven to be a lucrative design. The polymer membrane was specifically configured for this application by selectively engineering the size of the pores in the membrane to prevent air from passing, while allowing moisture to pass through the membrane. Polymer membranes can be designed to selectively allow particles of one size and shape to pass through while preventing others of different dimensions. This makes for a powerful tool that can be used in consumer products from biological weapons protection to industrial chemical separations.

A New York based company called Applied NanoWorks, Inc. has been developing a consumer product that utilizes LED technology to generate light. Light Emitting Diodes or LEDs, use only about 10% of the energy that a typical incandescent or fluorescent light bulb use and typically lasts much longer, which makes them a viable alternative to traditional light bulbs. While

LEDs have been around for decades, this company and others like it have been developing a special variant of LED called the white LED. White LEDs consist of semiconducting organic layers that are only about 100 nanometers in distance from each other and are placed between two electrodes, which create an anode, and a cathode.

When voltage is applied to the system, light is generated when electricity passes through the two organic layers. This is called electroluminescence. The semiconductor properties of the organic layers are what allow for the minimal amount of energy necessary to generate light. In traditional light bulbs, a metal filament is used to generate light when electricity is run through the filament. Using metal generates a great deal of heat and therefore lowers efficiency.

Research for longer lasting batteries has been an ongoing process for years. Researchers have now begun to utilize Nanotechnology for battery technology. mPhase Technologies in conglomeration with Rutgers University and Bell Laboratories have utilized nanomaterials to alter the Wetting Behavior of the surface where the liquid in the battery lies to spread the liquid droplets over a greater area on the surface and therefore have greater control over the movement of the droplets. This gives more control to the designer of the battery. This control prevents reactions in the battery by separating the electrolytic liquid from the anode and the cathode when the battery is not in use and joining them when the battery is in need of use.

ECONOMIC BENEFITS

The relatively recent shift toward using Nanotechnology with respect to the capture, transfer, and storage of energy has and will continue to have many positive economic impacts on society. The control of materials that Nanotechnology offers to scientists and engineers of consumer products is one of the most important aspects of Nanotechnology. This allows for an improved efficiency of products across the board.

A major issue with current energy generation is the loss of efficiency from the generation of heat as a by-product of the process. A common example of this is the heat generated by the internal combustion engine. The internal combustion engine losses about 36% of the energy from gasoline as heat and an improvement of this alone could have a significant economic impact. However, improving the internal combustion engine in this respect has proven to be extremely difficult without sacrificing performance. Improving the efficiency of fuel cells through the use of Nanotechnology appears to be more plausible by using molecularly tailored catalysts, polymer membranes, and improved fuel storage.

In order for a fuel cell to operate, particularly of the hydrogen variant, a noble-metal catalyst (usually platinum, which is very expensive) is needed to separate the electrons from the protons of the hydrogen atoms. However, catalysts of this type are extremely sensitive to carbon monoxide reactions. In order to combat this, alcohols or hydrocarbons compounds are used to lower the carbon monoxide concentration in the system. This adds an additional cost to the device. Using nanotechnology, catalysts can be designed through nanofabrication that are much more resistant to carbon monoxide reactions, which improves the efficiency of the process and may be designed with cheaper materials to additionally lower costs.

Fuel cells that are currently designed for transportation need rapid start-up periods for the practicality of consumer use. This process puts a lot of strain on the traditional polymer electrolyte membranes, which decreases the life of the membrane requiring frequent replacement. Using Nanotechnology, engineers have the ability to create a much more durable polymer membrane, which addresses this problem. Nanoscale polymer membranes are also much more efficient in ionic conductivity. This improves the efficiency of the system and decreases the time between replacements, which lowers costs.

Another problem with contemporary fuel cells is the storage of the fuel. In the case of hydrogen fuel cells, storing the hydrogen in gaseous rather than liquid form improves the efficiency by 5%. However, the materials that we currently have available to us significantly limit fuel storage due to low stress tolerance and costs. Scientists have come up with an answer to this by using a nanoporous styrene material (which is a relatively inexpensive material) that when super-cooled to around-196oC, naturally holds on to hydrogen atoms and when heated again releases the hydrogen for use.

CAPACITORS

For decades, scientists and engineers have been attempting to make computers smaller and more efficient. A crucial component of computers are capacitors. A capacitor is a device that is made of a pair of electrodes separated by an insulator that each stores an opposite charge. A capacitor stores a charge when it is removed from the circuit that it is connected to; the charge is released when it is replaced back into the circuit. Capacitors have an advantage over batteries in that they release their charge much more quickly than a battery.

Traditional or foil capacitors are composed of thin metal conducting plates separated by an electrical insulator, which are then stacked or rolled and placed in a casing. The problem with a traditional capacitor such as this is that they limit how small an engineer can design a computer. Scientists and engineers have since turned to Nanotechnology for a solution to the problem.

Using nanotechnology, researchers developed what they call "ultracapacitors." An ultracapacitor is a general term that describes a capacitor that contains nanocomponents. Ultracapacitors are being researched heavily because of their high density interior, compact size, reliability, and high capacitance. This decrease in size makes it increasingly possible to develop much smaller circuits and computers.

Ultracapacitors also have the capability to supplement batteries in hybrid vehicles by providing a large amount of energy during peak acceleration and allowing the battery to supply energy over longer periods of time, such as during a constant driving speed. This could decrease the size and weight of the large batteries needed in hybrid vehicles as well as take additional stress off the battery. However, as of now, the combination of ultracapacitors and a battery is not cost affective due to the need of additional DC/DC electronics to coordinate the two.

Nanoporous carbon aerogel is one type of material that is being utilized for the design of ultracapacitors. These aerogels have a very large interior surface area and can have its properties altered by changing the pore diameter and distribution along with adding nanosized alkali metals to alter its conductivity.

Carbon nanotubes are another possible material for use in an ultracapacitor. Carbon nanotubes are created by vaporizing carbon and allowing it to condense on a surface. When the carbon condenses, it forms a nanosized tube composed of carbon atoms. This tube has a high surface area, which increases the amount of charge that can be stored. The low reliability and high cost of using carbon nanotubes for ultracapacitors is currently an issue of research.

THEORY OF CAPACITANCE

Understanding the concept of capacitance can be helpful in understanding why Nanotechnology is such a powerful tool for the design of higher energy storing capacitors. A capacitor's capacitance (C) or amount of energy stored is equal to the amount of charge (Q) stored on each plate divided by the voltage (V) between the plates. Another representation of capacitance is that capacitance (C) is approximately equal to the permittivity (e) of the dielectric times the area (A) of the plates divided by the distance (d) between them. Therefore, capacitance is proportional to the surface area of the conducting plate and inversely proportional to the distance between the plates.

Using carbon nanotubes as an example, a property of carbon nanotubes is that they have a very high surface area to store a charge. Using the above proportionality that capacitance (C) is proportional to the surface area (A) of the conducting plate; it becomes obvious that using nanoscaled materials with high surface area would be great for increasing capacitance. The other proportionality described above is that capacitance (C) is inversely proportional to the distance (d) between the plates. Using nanoscaled plates such as carbon nanotubes with nanofabrication techniques, gives the capability of decreasing the space between plates which again increases capacitance.

Purpose

The Center on Nanotechnology and Society (Nano & Society) was created to catalyze informed interdisciplinary research, education, and dialogue on the ethical, legal, economic, policy, and broader societal implications of nanoscale science and technology.

As an affiliate of the Institute on Biotechnology and the Human Future, Nano & Society approaches nanotechnology with a special focus on the human condition. Nano & Society brings the foremost scholars and researchers in nano science, law, ethics, medicine, and the social sciences together with leaders in business and industry. Nano & Society also employs a number of vehicles in its efforts to focus a national conversation on nanotechnology and human dignity, including its website, the Chicago Nano Forum, national conferences, and NELSI Global (a public posicy document archive).

Online Offerings

Nano & Society conducts research on NELSI--Nanotechnology's ethical, legal, and social implications. The Center features various initiatvies: NELSI Global is a web-based global public policy document archive focused on nanotechnology's ethical, legal and societal implications (NELSI). This initiative is accessible through Nano & Society's website. NELSI Global serves as a unified clearinghouse, enabling users to identify and access key public policy documents addressing the NELSI agenda. It includes international and U.S.-specific regulations, legislation, case law, congressional testimony, and key governmental reports.

Publications, Conferences, and Events

The Center on Nanotechnology and Society uses multiple communications vehicles to educate and disseminate information on NELSI. These mechanisms include:

Nano and Society: The Center on Nanotechnology and Society publishes a monthly e-newsletter, entitled Nano & Society. which features opinion pieces from guest writers on nanotechnology's potential and potential impact, as well as public attitudes toward and education about nanotechnology; and relevant news updates on nanotechnology.

Nanologues: The Center on Nanotechnology and Society publishes a print series of booklets on NELSI entitled Nanologues.

Conferences and Events: Nano & Society hosts various national events to advance the dialogue on nanotechnology and the human condition, including:

Chicago Nano Forum: In conjunction with its partners in the educational, business, environmental, legal, and other arenas, Nano & Society has developed the Chicago Nano Forum. The Forum encourages public dialogue among nano experts in science, business, social sciences, ethics, and law. This series of events is webcast on the Center's website.

1st Annual Conference on Nanopolicy and the Human Future: This Nano & Society conference provided members of Congress and their staff, researchers, scientists, and others interested in and involved with the nano sector with the latest nanotechnology developments in the ethical, legal and social arena.

Center for Responsible Nanotechnology

The Center for Responsible Nanotechnology (CRN), founded in December, 2002, is a non-profit research and advocacy organization with a focus on molecular manufacturing and its possible effects, both positive and negative. CRN provides information to journalists, business leaders, policymakers and the general public about the environmental, humanitarian, economic, military, political, social, medical, and ethical implications of advanced nanotechnology.

CRN is an affiliate of World Care, an international non-profit 501(c)(3) organization. As of 2006, CRN's Director of Research is Chris Phoenix, and its Executive Director is Mike Treder. In August 2005, a task force of more than sixty international experts from various fields was organized by CRN to develop comprehensive recommendations for the safe and responsible use of nanotechnology. A series of eleven essays written by members of this CRN Task Force was published in March 2006 by the journal Nanotechnology Perceptions. The Center has sometimes been criticised for concentrating too heavily on the dangers posed by nanotechnology.

National Institute for Nanotechnology

The National Institute for Nanotechnology (NINT) is a Canadian research institution located on the University of Alberta main campus, in Edmonton, Alberta, Canada. Its primary purpose is nanotechnological research.

The institute was established in 2001 as a partnership between the National Research Council of Canada and the University of Alberta, and is funded by the Government of Canada, the Government of Alberta and the university. In June 2006, the institute moved into their present 20,000 square metre facility, designed to be one of the world's largest buildings for nanotechnological research. There are at most two or three other facilities worldwide matching the new building in scale and capacity. According to NINT's research plan, the institute's research focus is in the following areas:

- Synthesis and characterization of nanocrystals and nanowires
- Synthesis of supramolecular-based nanomaterials
- Fabrication and characterization of molecular-scale devices and nanosensors

- o Development of nano-scaled materials for catalysis and directed chemical reactions at semiconductor surfaces
- o Development of nano-electronic and nano-fluidic systems to interface devices to the outside world
- o Theory, modelling, and simulation of nanosystems on multiple length scales
- o Development of quantitative imaging and characterization techniques that support nanotechnology research

The work is focused around four interdisciplinary research groups:

- o Molecular scale development
- o Supramolecular nanoscale assembly
- o Materials and interfacial chemistry
- o Theory and modelling

COMPLEX NANOMATERIALS

To understand complex nanomaterials or nanomaterials having complex composition and their utility, a short narrative on nanomaterials is provided. Nanomaterials are materials possessing one or more dimensional features having a length on the order of a billionth of a meter to less than 100 billionths of a meter. They are important because: (1) they exhibit unique properties which are derived from the size of these features and (2) we have fairly recently learned how to manipulate matter on these dimensions so as to understand and exploit their unique properties and their relationship to size.

The properties of matter are largely a result of their composition, and the conditions under which they were produced.

The properties of most materials are largely a result of variations in composition, temperature and pressure and the rates at which these independent parameters are changed during their synthesis or production. As nanotechnologists, we look at "size" as an independent degree of freedom which can be manipulated independent of composition, temperature and pressure to yield materials that possess new properties not exhibited by their conventional counterparts. When materials possesses size features that are on the order of a few billionths of a meter, those materials often exhibit new properties not found in their ordinary material counterparts.

The NNI website which puts these ideas in a simple phrase, "At the nanoscale, the physical, chemical, and biological properties of materials differ in fundamental and valuable ways from the properties of individual atoms and molecules or bulk matter." This phrase can be further simplified to read: At the nanoscale, physical, chemical, and biological properties differ from the properties of individual atoms and molecules or bulk matter.

The nuances of how a material is produced often affects the molecular structure and the possible structure so produced give rise to variations in the materials bulk properties. An example of this is the case hardening of steel. A hard skin is created over a softer core by heating the steel to a high temperature and then rapidly cooling (quenching) it. The resulting composite structure, hard skin/softer core that develops is a function of the rate at which the temperature changed on the surface as compared to that in the core. (This will be discussed in greater detail below.)

To understand nanomaterials, normal materials (non-nanomaterials) are described to provide context. Materials are composed of atoms, molecules, ions, compositions thereof and assemblies of other materials. Solid materials can be broadly classified into crystalline and amorphous.

Crystalline materials come in two main varieties, single crystalline and polycrystalline. A crystalline material is composed of an orderly array of atoms, molecules or ions. Crystalline materials generally have short and long range order. That means that the manner in which atoms are arranged at any one location within a crystal is identical to the arrangement of at any other location. An example of a crystalline material is ordinary table salt. Each grain of salt is a single crystal. A diamond is also a single crystal. The semiconductor or chip inside a computer's microprocessor is a single crystal of silicon. An example of a polycrystalline material is a ceramic dish or coffee cup. If the broken edge of such item is examined with a magnifying glass one can easily see the individual crystals, referred to as crystallites or grains.

When a metal breaks, the grains or crystallites of which it is composed can be observed. In a figure coming soon is a metallograph of a polished metal surface. The grains are easily observed. The surface between grains is called the "grain boundary," and they appear as the line between grains in Figure coming soon. All crystalline materials exhibit Bragg scattering of x-rays. Amorphous materials are non crystalline. They have short range order but lack long range order. Polymers such as polyethylene, polyester and polypropylene are amorphous. Glass is an amorphous solid, which has short range order (always a Si next to an O, etc.) but lacks long range order. Amorphous materials are usually not composed of grains.

Nanomaterials include materials where the size of the particles, crystallites or grains of which the material is composed are on the order of nanometers. As a result of this properties emerge that are not characteristic of their ordinary counterparts having the same composition. These properties (examples given below) some from the fact that within each grain there is a relatively small number of atoms, molecules or ions as compared to a normal material. As the particle or grain size is in the nanometer range, many of the atoms or each grain or within the material as a whole reside at the particle surface, or grain boundary. The number of particle that reside at the surface or grain boundary is proportional to 1/r where r is the average radius of the particle size or grain.

That means that, as the particle size becomes smaller the number of atoms, molecules or ions that exist at the surface or grain boundary increases. Accordingly, the properties of a large collection of nanomaterials are dominated by the properties of the grain boundary or surface. As an example of how the size affects properties consider a material that exhibits ordered electronic spin on the surface but disordered electronic spin in the interior of the particle.

As the particle is reduced in size, the number of atoms on the surface increases as compared to those in the interior and so the number of ordered electronic spins increases as the particle size decreases. A material completely composed of nanoparticles or nano-crystallites of this type would exhibit properties of derived from the ordered spins, e.g. magnetism, while a normal material having the same composition would not exhibit such a property.

Nanomaterials can exist in several forms. The simplest are nanopowders. Nanopowders can be consolidated so as to form a polycrystalline nanomaterials. They can also be combined with other materials to form composite materials.

COMPARATIVE ANALYSIS

Complex materials are analogous to steel or other metal alloys because of their enhanced performance characteristics, while simple materials are more analogous to pure iron, aluminum or copper. An alloy is a mixture of several elements such as in the case of stainless steel: iron, nickel and chromium, each component of which imparts important characteristics to the performance of the alloy. A pure metal is predominantly composed of a single element, for example iron, as in an iron skillet or heating radiator found in homes built in the first half of the 20th century in this country. The first train rails were made out of iron, while all modern rails use steel alloys. Why? Because modern rails last longer, carry many times more weight, are less susceptible to corrosion, and will not become brittle and crack with changing temperatures. Virtually all modern metals in ordinary use are made out of alloys that improve performance, strength and longevity.

Early successes in the nanomaterials industry (use of nano particles in sun screen and ultra fine polishing of semiconductor wafers) represent simple material applications and in these applications only rudimentary aspects of nanotechnology are exploited. Based on current performance limitations, future developments in fuel cells, solar technology, electronics, optics, biotechnology, cosmetics and advanced structural materials involve solutions requiring complex nanomaterials.

NANOTOXICOLOGY

Nanotoxicology is the study of the toxicity of nanomaterials. Because of the small size and large surface area of nanomaterials, these materials have unique properties compared with their larger counterparts. The nanomaterials, even when they are made of inert elements like gold, become very active at a nanometer range. Nanotoxicological studies are intended to determine whether and to what extent these may pose a threat to the environment and to human beings.

Research on ultrafine particles has laid the foundation for the emerging field of nanotoxicology, with the goal of studying the biokinetics of engineered nanomaterials and their potential for causing adverse effects. Most reports find that ultrafine particles are more toxic than equivalent larger-sized particles of a given material at similar doses per gram of body weight.

Assessing the safety of engineered nanoparticles is a highly complex matter that goes beyond traditional toxicology. Engineered nanoparticles are not a uniform group of substances. The problem arises from the fact that particle size alone (which determines the surface area of a given mass of a substance) is not the only factor that determines the toxicological impact of a material and that makes nanoparticles of a given substance behave differently from the bulk form.

While the release and production of nanoparticles during industrial and combustion processes and activities is mostly unintentional, the emergence of nanotechnological production processes introduces the intentional and controlled manufacture of nanoparticles. The latter can be further differentiated as either bulk nanoparticles in industry, e.g., carbon black or titanium dioxide, or so-called engineered nanoparticles, e.g., carbon nanotubes.

Given the prospects for nanotechnology, and the fact that products containing engineered nanoparticles have already been introduced to the marketplace, the increasing flow of new products will bring about the massive production of engineered nanoparticles.

Despite all the scientific knowledge gained in the toxicology of particulate matter, because of the many variables involved, scientists still cannot accurately predict how nanomaterials will affect living organisms. What is clear, though, is that the biologic activity and biokinetics of nanoparticles are different from larger particles, and that they depend on many parameters. These parameters can modify cellular uptake, protein binding, translocation from portal of entry to target site, and the possibility of tissue injury.

Important from a toxicological point of view are the physiochemical properties that come with size and lead to certain biological reactivity. Size, together with differences in shape, surface structure, chemical composition (purity, crystallinity, electrical properties, etc.), solubility and biopersistence make for a large number of variables that need to be considered when assessing nanoparticles. On top of that, powders or liquids containing nanoparticles are almost never monodisperse but contain a range of particle sizes. This complicates experimental analysis as larger nanoparticles might have different properties than smaller ones. Also, nanoparticles show a tendency to aggregate and such aggregates often behave differently from individual nanoparticles.

Most, if not all, toxicological studies on nanoparticles rely on current methods, practices and terminology as gained and applied in the analysis of micro-and ultrafine particles and mineral fibers. Together with recent studies on nanoparticles, this provides an initial basis for evaluating the primary issues in a risk assessment framework for nanomaterials.

Given the many parameters involved, nanotoxicology requires an interdisciplinary team approach, even more so than classical toxicology, in order to arrive at an appropriate risk assessment. As a still-maturing science, nanotoxicology will expand the boundaries of traditional toxicology from a testing and auxiliary science to a new discipline where toxicological knowledge of nanomaterials can be put to constructive use in therapeutics as well as the development of new and better biocompatible materials.

Nanotribology

Nanotribology is a branch of tribology which studies friction phenomenon at the nanometer scale. The distinction between nanotribology and tribology is primarily due to the involvement of atomic forces in the determination of the final behavior of the system.

Gears, bearings, and liquid lubricants can reduce friction in the macroscopic world, but the origins of friction for small devices such as micro-or nanoelectromechanical systems (NEMS) require other solutions. Despite the unprecedented accuracy by which these devices are nowadays designed and fabricated, their enormous surface-volume ratio leads to severe friction and wear issues, which dramatically reduce their applicability and lifetime. Traditional liquid lubricants become too viscous when confined in layers of molecular thickness. This situation has led to a number of proposals for ways to reduce friction on the nanoscale, such as superlubricity and thermolubricity.

Nanophotonics

Nanophotonics is the study of the behavior of light on the nanometre scale. The ability to fabricate devices in nanoscale that has been developed recently provided the catalyst for this area of study.

Nanophotonics scientists are experimenting with different ways to generate and manipulate light using ultrasmall, engineered structures at the nanoscale. Tiny gold particles called "nanostars" are being studied to ascertain how they interact with light. The study has the potential to revolutionise the telecommunications industry by providing low power, high speed, interference-free devices such as electrooptic and all-optical switches on a chip. There are a number of research groups working in this area in the U.S., the UK, Italy and Spain.

Mechanochemistry

Mechanochemistry, is the coupling of the mechanical and the chemical on a molecular scale includes mechanical breakage, polymer degradation under shear, cavitation related phenomena (e.g., sonochemistry and sonoluminescence), shockwave chemistry and physics, and even the burgeoning field of molecular machines.

A small part of mechanochemistry is sometimes also called "positional synthesis" or "positional assembly" is a technique for forming chemical bonds by direct computer control of the position of molecules. This is an example of a specific type of Mechanosynthesis.

As of 2004, the typical experimental arrangement is to attach a molecule to the tip of an atomic force microscope, and then use the microscope's precise positioning abilities to push the molecule on the tip into another on a substrate. Since the angles and distances can be precisely controlled, and the reaction occurs in a vacuum, novel chemical compounds and arrangements are possible. Much of the excitement regarding mechanochemistry regards its potential use in automated assembly of molecular-scale devices. Such techniques appear to have many applications in medicine, aviation, resource extraction, manufacturing and warfare.

Most theoretical explorations of such machines have focused on using Carbon, because of the many **strong** bonds it can form, the many types of chemistry these bonds permit, and utility of these bonds in medical and mechanical applications. Carbon forms diamond, for example, which if cheaply available, would be an excellent material for many machines. In practice, getting exactly one molecule to a known place on the microscope's tip is possible, but has proven difficult to automate. Since practical products require at least several hundred million atoms, this technique has not yet proven practical in forming a real product.

The goal of mechanoassembly research at this point focuses on overcoming these problems by calibration, and selection of appropriate synthesis reactions. The first product to be built by these means will probably be a specialized, very small (roughly 1,000 nanometers on a side) machine tool that can build copies of itself using mechanochemical means, under the control of an external computer. In the literature, such a tool is called an assembler. Once assemblers exist, geometric growth (copies making copies) could reduce the cost of assemblers rapidly. Control by an external computer should then permit large groups of assemblers to construct large, useful projects to atomic precisions. One such project would combine molecular-level conveyor belts with permanently-mounted assemblers to produce a factory.

Mechanochemistry actually goes back more than 100 years. The coupling of the mechanical and the chemical on a molecular scale includes mechanical breakage, polymer degradation under shear, cavitation related phenomena (e.g., sonochemistry and sonoluminescence), shockwave chemistry and physics, and even the burgeoning field of molecular machines.

The technique of moving single atoms mechanically was proposed by Eric Drexler in his 1986 book The Engines of Creation. In 1988, researchers at IBM's Zürich Research Institute successfully spelled the letters "IBM" in Xenon atoms on a cryogenic copper surface, grossly validating the approach.

Since then, a number of research projects have undertaken to use similar techniques to store computer data in a compact fashion. More recently the technique has been used to explore novel physical chemistries, sometimes using lasers to excite the tips to particular energy states, or examine the quantum chemistry of particular chemical bonds.

Nano-optics

Nano-optics is the branch of optical engineering which deals with optics at deeply subwavelength length scales. Technologies in the realm of nano-optics include near-field scanning optical microscopy (NSOM), photoassisted scanning tunnelling microscopy, and surface plasmon optics. Traditional microscopy makes use of diffractive elements to focus light tightly in order to increase resolution.

But because of the diffraction limit (also known as the Rayleigh Criterion), propagating light may be focused to a spot with a minimum diameter of roughly half the wavelength of light. Thus, even with diffraction-limited confocal microscopy, the maximum resolution obtainable is on the order of a couple hundred nanometers. The scientific and industrial communities are becoming more interested in the characterization of materials and phenomena on the scale of a few nanometers, so alternative techniques must be utilized. Scanning Probe Microscopy (SPM) makes use of a "probe," (usually either a tiny aperture or super-sharp tip), which either locally excites a sample or transmits local information from a sample to be collected and analyzed.

Nanolanguage

NanoLanguage is a new way of thinking scientific computing, combining the strength of flexible object-oriented scripting interfaces (known from Mathematica and MatLab) with sophisticated high performance scientific computing algorithms. The goal is to enable scientists to efficiently extend, specialize and combine methods to calculate nanoscale properties of matter, including density functional theory, semi-empirical tight-binding, classical potentials, k.p and various quantum-chemical methods. NanoLanguage allows for both low and high level detailed control of the computer simulations performed in the Atomistix ToolKit. At the high level, it offers a common interface for setting up complex atomic-scale simulations and analyzing the results. On the lower level side, it provides an interface to the low-level functionality in ATK.

NanoLanguage is built on top of Python, a powerful and well-established interpreted programming language, and thus it includes basic elements such as loops over simulation control parameters, plus support for efficient manipulations of e.g. numerical array data. It is therefore an ideal tool for automating series of simulations where geometric, material, or other parameters are to be optimized.

Using NanoLanguage allows scientists to express models of nature in a common language without the need to re-implement already implemented algorithms, and it allows for third-party development of new functionality on top of the ATK platform. Such functionality may consist of new atomic-scale modeling methodologies, tailored semi-empirical methods, or complex post-processing methods for calculating new quantities from the fundamental simulation results.

Nanocrystal Solar Cell

Nanocrystal solar cells or quantum dot solar cells, are solar cells based on a silicon substrate with a coating of nanocrystals. Whilst previous methods of quantum dot creation relied on expensive molecular beam epitaxy processes, fabrication using colloidal synthesis allows for a more cost effective manufacture. A thin film of nanocrystals is obtained by a process known as "spin-coating". This involves placing an amount of the quantum dot solution onto a flat substrate, which is then rotated very quickly. The solution spreads out uniformly, and the substrate is spun until the required thickness is achieved.

Quantum dot based photovoltaic cells based around dye-sensitised colloidal TiO2 films were investigated in 1991 and were found to exhibit promising efficiency of converting incident light energy to electrical energy, and were found to be incredibly encouraging due to the low cost of materials in the search for more commercially viable/affordable renewable energy sources.

Although research is still in its infancy and is ongoing, in the future quantum dot based photovoltaics may offer advantages such as mechanical flexiblity (quantum dot-polymer composite photovoltaics) as well as low cost, clean power generation.

11

Nanomanufacturing and NanoMEMS

Nanomanufacturing is used to describe either the production of nano scaled materials, which can be powders or fluids, or describe the manufacturing of parts "button up" from nanoscaled materials or "top down" in smallest steps for high precison by several technologis like laser ablation, etching and others. The term is widely used, e.g. by the European Technology Platform MINAM and the U.S. National Nanotechnology Initiative (NNI). NNI refer to a sub-effort of nanotechnology--one of its five "priority areas."

There is also a Nanomanufacturing Program at the U.S. National Science Foundation. Nanomanufacturing should not be confused with molecular manufacturing which refers specifically to the manufacture of complex, nanoscale structures by means of nonbiological mechanosynthesis (and subsequent assembly). The NNI has defined nanotechnology very broadly, to include a wide range of tiny structures including those created by large and imprecise tools.

Nanomanufacturing is not defined in the NNI's recent report, "Instrumentation and Metrology for Nanotechnology." (For contrast, another "priority area," nanofabrication, is defined as "the ability to fabricate, by directed or self-assembly methods, functional structures or devices at the atomic or molecular level." p. 67) Nanomanufacturing appears to be the near-term industrial-scale manufacture of nanotechnology-based objects, with emphasis on low cost and reliability.

THE CATALYSTS

Nanomaterial-based catalysts are usually heterogeneous catalysts broken up into nanoparticles in order to speed up the catalytic process. The extremely small size of the particles maximizes the surface area exposed to the reactant, allowing more reactions to occur at the same time, thus speeding up the process. Much research on nanomaterial-based catalysts have to do with maximizing the effectiveness of the catalyst coating in fuel cells. Platinum is currently the most common catalyst for this application, however, it is expensive and rare, so a lot of research has been going into maximizing the catalytic properties of other metals by shrinking them to nanoparticles in the hope that someday they will be an efficient and economic alternative to platinum.

Defense programs in many countries are now concentrating on nanotechnology research that will facilitate advances in such technology used to create secure but small messaging equipment, allow the development of smart weapons, improve stealth capabilities, aid in developing specialized sensors (including bio-inclusive sensors), help to create self-repairing military equipment, and improve the development and delivery mechanisms for medicines and vaccines. Nanotechnology builds on advances in microelectronics during the last decades of the twentieth century. The miniaturization of electrical components greatly increased the utility and portability of computers, imaging equipment, microphones, and other electronics.

Indeed, the production and wide use of such commonplace devices such as personal computers and cell phones was absolutely dependent on advances in microtechnology.

Despite these fundamental advances there remain real physical constraints (e.g., microchip design limitations) to further miniaturization based upon conventional engineering principles. Nanotechnologies intend to revolutionize components and manufacturing techniques to overcome these fundamental limitations. In addition, there are classes of biosensors and feedback control devices that require nanotechnology because-despite advances in microtechnology-present components remain too large or slow.

ADVANCES

Nanotechnology advances affect all branches of engineering and science that deal directly with device components ranging in size between 1/10,000,000 (one ten millionth of a millimeter) and 1/10,0000 millimeter. At these scales, even the most sophisticated microtechnology-based instrumentation is useless. Engineers anticipate that advances in nanotechnology will allow the direct manipulation of molecules in biological samples (e.g., proteins or nucleic acids) paving the way for the development of new materials that have a biological component or that can provide a biological interface.

In addition to new tools, nanotechnology programs advance practical understanding of quantum physics. The internalization of quantum concepts is a necessary component of nanotechnology research programs because the laws of classical physics (e.g., classical mechanics or generalized gas laws) do not always apply to the atomic and near-atomic level. Nanotechnology and quantum physics. Quantum theory and mechanics describe the relationship between energy and matter on the atomic and subatomic scale. At the beginning of the twentieth century, German physicist Maxwell Planck (1858-1947) proposed that atoms absorb or emit electromagnetic radiation in bundles of energy termed quanta. This quantum concept seemed counter-intuitive to well-established Newtonian physics. Advancements associated with quantum mechanics (e.g., the uncertainty principle) also had profound implications with regard to the philosophical scientific arguments regarding the limitations of human knowledge.

Planck's quantum theory, which also asserted that the energy of light (a photon) was directly proportional to its frequency, proved a powerful concept that accounted for a wide range of physical phenomena. Planck's constant relates the energy of a photon with the frequency of light. Along with the constant for the speed of light, Planck's constant (h = 6.626 x 10-34 Joule-second) is a fundamental constant of nature.

Prior to Planck's work, electromagnetic radiation (light) was thought to travel in waves with an infinite number of available frequencies and wavelengths. Planck's work focused on attempting to explain the limited spectrum of light emitted by hot objects. Danish physicist Niels Bohr (1885-1962) studied Planck's quantum theory of radiation and worked in England with physicists J. J. Thomson (1856-1940), and Ernest Rutherford (1871-1937) to improve their classical models of the atom by incorporating quantum theory. During this time, Bohr developed his model of atomic structure.

According to the Bohr model, when an electron is excited by energy it jumps from its ground state to an excited state (i.e., a higher energy orbital). The excited atom can then emit energy only in certain (quantized) amounts as its electrons jump back to lower energy orbits located closer to the nucleus. This excess energy is emitted in quanta of electromagnetic radiation (photons of light) that have exactly the same energy as the difference in energy between the orbits jumped by the electron.

The electron quantum leaps between orbits proposed by the Bohr model accounted for Plank's observations that atoms emit or absorb electromagnetic radiation in quanta. Bohr's model also explained many important properties of the photoelectric effect described by Albert Einstein (1879-1955). Einstein assumed that light was transmitted as a stream of particles termed photons. By extending the well-known wave properties of light to include a treatment of light as a stream of photons, Einstein was able to explain the photoelectric effect. Photoelectric properties are key to regulation of many microtechnology and proposed nanotechnology level systems.

Quantum mechanics ultimately replaced electron "orbitals" of earlier atomic models with allowable values for angular momentum (angular velocity multiplied by mass) and depicted electron positions in terms of probability "clouds" and regions.

In the 1920s, the concept of quantization and its application to physical phenomena was further advanced by more mathematically complex models based on the work of the French physicist Louis Victor de Broglie (1892-1987) and Austrian physicist Erwin Schrödinger (1887-1961) that depicted the particle and wave nature of electrons. De Broglie showed that the electron was not merely a particle but a waveform. This proposal led Schrödinger to publish his wave equation in 1926. Schrödinger's work described electrons as a "standing wave" surrounding the nucleus, and his system of quantum mechanics is called wave mechanics. German physicist

Max Born (1882-1970) and English physicist P. A. M. Dirac (1902-1984) made further advances in defining the subatomic particles (principally the electron) as a wave rather than as a particle and in reconciling portions of quantum theory with relativity theory.

Working at about the same time, German physicist Werner Heisenberg (1901-1976) formulated the first complete and self-consistent theory of quantum mechanics. Matrix mathematics was well established by the 1920s, and Heisenberg applied this powerful tool to quantum mechanics. In 1926, Heisenberg put forward his uncertainty principle which states that two complementary properties of a system, such as position and momentum, can never both be known exactly. This proposition helped cement the dual nature of particles (e.g., light can be described as having both wave and particle characteristics). Electromagnetic radiation (one region of the spectrum that comprises visible light) is now understood to have both particle and wave like properties.

In 1925, Austrian-born physicist Wolfgang Pauli (1900-1958) published the Pauli exclusion principle states that no two electrons in an atom can simultaneously occupy the same quantum state (i.e., energy state). Pauli's specification of spin (+1/2 or-1/2) on an electron gave the two electrons in any suborbital differing quantum numbers (a system used to describe the quantum state) and made completely understandable the structure of the periodic table in terms of electron configurations (i.e., the energy-related arrangement of electrons in energy shells and suborbitals).

In 1931, American chemist Linus Pauling published a paper that used quantum mechanics to explain how two electrons, from two different atoms, are shared to make a covalent bond between the two atoms. Pauling's work provided the connection needed in order to fully apply the new quantum theory to chemical reactions. Advances in nanotechnology depend upon an understanding and application of these fundamental quantum principles. At the quantum level the smoothness of classical physics disappears and nanotechnologies are predicated on exploiting this quantum roughness.

APPLICATIONS

The development of devices that are small, light, self-contained, use little energy and that will replace larger microelectronic equipment is one of the first goals of the anticipated nanotechnology revolution. The second phase will be marked by the introduction of materials not feasible at larger than nanotechnology levels. Given the nature of quantum variance, scientists theorize that single molecule sensors can be developed and that sophisticated memory storage and neural-like networks can be achieved with a very small number of molecules. Traditional engineering concepts undergo radical transformation at the atomic level. For example, nano-technology motors may drive gears, the cogs of which are composed of the atoms attached to a carbon ring. Nanomotors may themselves be driven by oscillating magnetic fields or high precision oscillating lasers.

Perhaps the greatest promise for nanotechnology lies in potential biotechnology advances. Potential nano-level manipulation of DNA offers the opportunity to radically expand the horizons of genomic medicine and immunology. Tissue-based biosensors may unobtrusively be able to monitor and regulate site-specific medicine delivery or regulate physiological processes. Nanosystems might serve as highly sensitive detectors of toxic substances or used by inspectors to detect traces of biological or chemical weapons.

In electronics and computer science, scientists assert that nanotechnologies will be the next major advance in computing and information processing science. Microelectronic devices rely on recognition and flips in electron gating (e.g. where differential states are ultimately represented by a series of binary numbers ["0" or "1"] that depict voltage states).

In contrast, future quantum processing will utilize the identity of quantum states as set forth by quantum numbers. In quantum cryptography systems with the ability to decipher encrypted information will rely on precise knowledge of manipulations used to achieve various atomic states.

Nanoscale devices are constructed using a combination of fabrication steps. In the initial growth stage, layers of semiconductor materials are grown on a dimension limiting substrate. Layer composition can be altered to control electrical and/or optical characteristics. Techniques such as molecular beam epitaxy (MBE) and metallo-organic chemical vapor deposition (MOCVD) are capable of producing layers of a few atoms thickness. The developed pattern is then imposed on successive layers (the pattern transfer stage) to develop desired three dimensional structural characteristics.

NANOTECHNOLOGY RESEARCH

In the United States, expenditures on nanotechnology development tops $500 million per year and is largely coordinated by the National Science Foundation and Department of Defense Advanced Research Projects Agency (DARPA) under the umbrella of the National Nano-technology Initiative. Other institutions with dedicated funding for nanotechnology include the Department of Energy (DOE) and National Institutes of Health (NIH).

Research interests. Current research interests in nanotechnology include programs to develop and exploit nanotubes for their ability to provide extremely strong bonds. Nanotubes can be flexed and woven into fibers for use in ultrastrong-but also ultralight-bulletproof vests. Nanotubes are also excellent conductors that can be used to develop precise electronic circuitry.

Other interests include the development of nanotechnology-based sensors that allow smarter autonomous weapons capable of a greater range of adaptations enroute to a target; materials that offer stealth characteristics across a broader span of the electromagnetic spectrum; self-repairing structures; and nanotechnology-based weapons to disrupt-but not destroy-electrical system infrastructure.

12

Progress of Nanotechnology

A basic question about nanotechnology is, "When will it be achieved?" The answer is simple: No one knows. How molecular machines will behave is a matter for calculation, but how long it will take us to develop them is a separate issue. Technology timetables can't be calculated from the laws of nature, they can only be guessed at. In this chapter, we examine different paths to nanotechnology, hear what some of the pioneers have to say, and describe the progress already made. This will not answer our basic question, but it will educate our guesses.

Molecular nanotechnology could be developed in any of several basically different ways. Each of these basic alternatives itself includes further alternatives. Researchers will be asking, "How can we make the fastest progress?" To understand the answers they may come to, we need to ask the same question here, adopting (for the moment) a gung-ho, let's-go, how-do-we-get-the-job-done? attitude. We give some of the researchers' answers in their own words.

Like "When will it be achieved?", this is a basic question with an answer beyond calculation. Here, though, the answer seems fairly clear. Throughout history, people have worked to achieve better control of matter, to convince atoms to do what we want them to do. This has gone on since before people learned that atoms exist, and has accelerated ever since. Although different industries use different materials and different tools and methods, the basic aim is always the same. They seek to make better things, and make them more consistently, and that means better control of the structure of matter. From this perspective, nanotechnology is just the next, natural step in a progression that has been under way for millennia.

Consider the compact discs now replacing older stereo records: both the old and the new technologies stamp patterns into plastic, but for CDs, the bumps on the stamping surface are only about 130 by 600 nanometers in size, versus 100,000 nanometers or so for the width of the groove on an old-style record. Or look at a personal computer.

John Foster, a physicist at IBM's Almaden Research Center, points to a hard disk and says that "inside that box are a bunch of whirring disks, and every one of those disks has got a metal layer where the information is stored. The last thing on top of the metal layer is a monolayer that's the lubricant between the disk and the head that flies over it. The monolayer is not fifteen angstroms [15 angstroms = 1.5 nanometers] and it's not three, because fifteen won't work and neither will three. So it has to be ten plus or minus a few angstroms. This is definitely working in the nanometer regime. We're at that level: We ship it every day and make money on it every day."

The transistors on computer chips are heading down in size on an exponential curve. Foster's colleague at IBM, Patrick Arnett, expects the trend to continue: "If you stay on that curve, then you end up at the atomic scale at 2020 or so. That's the nature of technology now. You expect to follow that curve as far as you can go."

The trend is clear, and at least some of the results can be foreseen, but the precise path and timetable for the development of nanotechnology is unpredictable. This unpredictability goes to the heart of important questions: "How will this technology be developed? Who will do it? Where? When? In ten years? Fifty? A hundred? Will this happen in my lifetime?" The answers will depend on what people do with their time and resources, which in turn will depend on what goals they think are most promising. Human attitudes, understanding, and goals will make all the difference.

RATE OF ADVANCE

Decisions about research directions are central. Researchers are already pouring effort into chemical synthesis, molecular engineering, and related fields. The same amount of effort could produce more impressive results in molecular nanotechnology if a fraction of it were differently directed. The research funders-corporate executives, and decision makers in science funding agencies like the National Science Foundation in the United States and Japan's Ministry of International Trade and Industry-all have a large influence on research directions, but so do the researchers working in the labs. They submit proposals to potential funders (and often spend time on personally chosen projects, regardless of funding), so their opinions also shape what happens. Where public money is involved, politicians' impressions of public opinion can have a huge influence, and public opinion depends on what all of us think and say..

Still, researchers play a central role. They tend to work on what they think is interesting, which depends on what they think is possible, which depends on the tools they have or-among the most creative researchers-on the tools they can see how to make. Our tools shape how we think: as the saying goes, when all you have is a hammer, everything looks like a nail. New tools encourage new thoughts and enable new achievements, and decisions about tool development will pace advances in nanotechnology. To understand the challenges ahead, we need to take a look at ideas about the tools that will be needed.

Imporatance of Tools

Throughout history, limited tools have limited achievement. Leonardo da Vinci's sixteenth century chain drives and ball bearings were theoretically workable, yet never worked in their inventor's lifetime. Charles Babbage's nineteenth century mechanical computer suffered the same fate. The problem? Both inventors needed precisely machined parts that (though readily available today) were beyond the manufacturing technology of their times. Physicist David Miller recounts how a sophisticated integrated circuit design project at TRW hit similar limits in the early 1980s: "It all came down to whether a German company could cool their glass lenses slowly enough to give us the accuracy we needed. They couldn't."

In the molecular world, tool development again paces progress, and new tools can bring breathtaking advances. Mark Pearson, director of molecular biology for Du Pont, has observed this in action: "When I was a graduate student back in the 1950s, it was a multiyear problem to determine the molecular structure of a single protein. We used to say, 'one protein, one career.' Yet now the time has shrunk from a career to a decade to a year-and in optimal cases to a few months."

Protein structures can be mapped atom by atom by studying X-ray reflections from layers in protein crystals.

Pearson observes that "Characterizing a protein was a career-long endeavor in part because it was so difficult to get crystals, and just getting the material was a big constraint. With new technologies, we can get our hands on the material now-that may sound mundane, but it's a great advance. To the people in the field, it makes all the difference in the world." Improved tools for making and studying proteins are of special importance because proteins are promising building blocks for first-generation molecular machines.

SCIENCE DISCOVERY

Nobel Prizes are more often awarded for discoveries than for the tools (including instruments and techniques) that made them possible. If the goal is to spur scientific progress, this is a shame. This pattern of reward extends throughout science, leading to a chronic underinvestment in developing new tools. Philip Abelson, an editor of the journal Science, points out that the United States suffers from "a lack of support for development of new instrumentation. At one time, we had a virtual monopoly in pioneering advances in instrumentation. Now practically no federal funds are available to universities for the purpose." It's easier and less risky to squeeze one more piece of data out of an existing tool than to pioneer the development of a new one, and it takes less imagination.

But new tools emerge anyway, often from sources in other fields. The study of protein crystals, for example, can benefit from new X-ray sources developed by physicists, and techniques from chemistry can help make new proteins. Because they can't anticipate tools resulting from innovations in other fields, scientists and engineers are often too pessimistic about what can be achieved in their own fields. Nanotechnology will join several fields, and yield tools useful in many others. We should expect surprising results.

Researchers' Tools

Today's tools for making small-scale structures are of two kinds: molecular-processing tools and bulk-processing tools. For decades, chemists and molecular biologists have been using better and better molecular-processing tools to make and manipulate precise, molecular structures. These tools are of obvious use. Physicists, as we will see, have recently developed tools that can also manipulate molecules. Combined with techniques from chemistry and molecular biology, these physicist's tools promise great advances. Microtechnologists have applied chip-making techniques to the manufacture of microscopic machines. These technologies-the main approach to miniaturization in recent decades-can play at most a supporting role in the development of nanotechnology. Despite appearances, it seems that microtechnology cannot be refined into nanotechnology.

Small Microtechnology

For many years, it was conventional to assume that the road to very small devices led through smaller and smaller devices: a top-down path. On this path, progress is measured by miniaturization: How small a transistor can we build? How small a motor? How thin a line can we draw on the surface of a crystal? Miniaturization focuses on scale and has paid off well, spawning industries ranging from watchmaking to microelectronics.

Researchers at AT&T Bell Labs, the University of California at Berkeley, and other laboratories in the United States have used micromachining (based on microelectronic technologies) to make tiny gears and even electric motors. Micromachining is also being pursued successfully in Japan and Germany. These microgears and micromotors are, however, enormous by nanotechnological standards: a typical device is measured in tens of micrometers, billions of times the volume of comparable nanogears and nanomotors. (In our simulated molecular world, ten microns is the size of a small town.) In size, confusing microtechnology with molecular nanotechnology is like confusing an elephant with a ladybug.

The differences run deeper, though. Microtechnology dumps atoms on surfaces and digs them away again in bulk, with no regard for which atom goes where. Its methods are inherently crude. Molecular nanotechnology, in contrast, positions each atom with care. As Bill DeGrado, a protein chemist at Du Pont, says, "The essence of nanotechnology is that people have worked for years making things smaller and smaller until we're approaching molecular dimensions. At that point, one can't make smaller things except by starting with molecules and building them up into assemblies." The difference is basic: In microtechnology, the challenge is to build smaller; in nanotechnology, the challenge is to build bigger-we can already make small molecules.

A language warning: in recent years, nanotechnology has indeed been used to mean "very small microtechnology"; for this usage, the answer to the above question is yes, by definition. This use of a new word for a mere extension of an old technology will produce considerable confusion, particularly in light of the widespread use of nanotechnology in the sense found here. Nanolithography, nanoelectronics, nanocomposites, nanofabrication: not all that is nano- is molecular, or very relevant to the concerns raised in this book. The terms molecular nanotechnology and molecular manufacturing are more awkward but avoid this confusion.

Can bulldozers can be used to make wristwatches? At most, they can help to build factories in which watches are made. Though there could be surprises, the relevance of microtechnology to molecular nanotechnology seems similar. Instead, a bottom-up approach is needed to accomplish engineering goals on the molecular scale.

Tools Used for Molecular Engineering

Almost by definition, the path to molecular nanotechnology must lead through molecular engineering. Working in different disciplines, driven by different goals, researchers are making progress in this field. Chemists are developing techniques able to build precise molecular structures of sorts never before seen. Biochemists are learning to build structures of familiar kinds, such as proteins, to make new molecular objects.

In a visible sense, most of the tools used by chemists and biochemists are rather unimpressive. They work on countertops cluttered with dishes, bottles, tubes, and the like, mixing, stirring, heating, and pouring liquids-in biochemistry, the liquid is usually water with a trace of material dissolved in it. Periodically, a bit of liquid is put into a larger machine and a strip of paper comes out with a graph printed on it. As one might guess from this description, research in the molecular sciences is usually much less expensive than research in high-energy physics (with its multibillion-dollar particle accelerators) or research in space (with its multibillion-dollar spacecraft).

Chemistry has been called "small science," and not because of the size of the molecules. Of these fields, it is biomolecular science that is most obviously developing tools that can build nanotechnology, because biomolecules already form molecular machines, including devices resembling crude assemblers. This path is easiest to picture, and can surely work, yet there is no guarantee that it will be fastest: research groups following another path may well win. Each of these paths is being pursued worldwide, and on each, progress is accelerating.

Physicists have recently contributed new tools of great promise for molecular engineering. These are the proximal probes, including the scanning tunneling microscope (STM) and the atomic force microscope (AFM). A proximal-probe device places a sharp tip in proximity to a surface and uses it to probe (and sometimes modify) the surface and any molecules that may be stuck to it.

MOLECULAR BUILDING BLOCKS

A huge technology base for molecular construction already exists. Tools originally developed by biochemists and biotechnologists to deal with molecular machines found in nature can be redirected to make new molecular machines. The expertise built up by chemists in more than a century of steady progress will be crucial in molecular design and construction. Both disciplines routinely handle molecules by the billions and get them to form patterns by self-assembly. Biochemists, in particular, can begin by copying designs from nature.

Molecular building-block strategies could work together with proximal probe strategies, or could replace them, jumping directly to the construction of large numbers of molecular machines. Either way, protein molecules are likely to play a central role, as they do in nature.

Protein Engineering Building Molecular Machines

Proteins can self assemble into working molecular machines, objects that do something, such as cutting and splicing other molecules or making muscles contract. They also join with other molecules to form huge assemblies like the ribosome (about the size of a washing machine, in our simulation view). Ribosomes-programmable machines for manufacturing proteins-are nature's closest approach to a molecular assembler. The genetic-engineering industry is chiefly in the business of reprogramming natural nanomachines, the ribosomes, to make new proteins or to make familiar proteins more cheaply. Designing new proteins is termed protein engineering. Since biomolecules already form such complex devices, it's easy to see that advanced protein engineering could be used to build first-generation nanomachines.

Fancy Molecular Machines

Making proteins is easier than designing them. Protein chemists began by studying proteins found in nature, but have only recently moved on to the problem of engineering new ones. These are called de novo proteins, meaning completely new, made from scratch. Designing proteins is difficult because of the way they are constructed. As Bill DeGrado, a protein chemist at Du Pont, explains: "A characteristic of proteins is that their activities depend on their three-dimensional structures. These activities may range from hormonal action to a function in digestion or in metabolism.

Whatever their function, it's always essential to have a definite three-dimensional shape or structure." This three-dimensional structure forms when a chain folds to form a compact molecular object. To get a feel for how tough it is to predict the natural folding of a protein chain, picture a straight piece of cord with hundreds of magnets and sticky knots along its length. In this state, it's easy to make and easy to understand. Now pick it up, put it in a glass jar, and shake it for a long time. Could you predict its final shape? Certainly not: it's a tangled mess. One might call this effort at prediction "the sticky-cord-folding problem"; protein chemists call theirs "the protein-folding problem."

Given the correct conditions, a protein chain always folds into one special shape, but that shape is hard to predict from just the straightened structure. Protein designers, though, face the different job of first determining a desired final shape, and then figuring out what linear sequence of amino acids to use to make that shape. Without solving the classic protein-folding problem, they have begun to solve the protein-design problem.

The Accomplishments So Far

Bill DeGrado and his colleagues at Du Pont had one of the first successes: "We've been able to use basic principles to design and build a simple molecule that folds up the way we want it to. This is really the first real example of a designed protein structure, designed from scratch, not by taking an already existing structure and tinkering with it."

Although scientists do the work, the work itself is really a form of engineering, as shown by the title of the field's journal, Protein Engineering.

Bill DeGrado's description of the process makes this clear: "After you've made it, the next step is to find out whether your protein did what you expected it to do. Did it fold? Did it pass ions across bilayers [such as cell membranes]? Does it have a catalytic function [speeding specific chemical reactions]? And that's tested using the appropriate experiment. More than likely, it won't have done what you wanted it to do, so you have to find out why. Now, a good design has in it a contingency plan for failure and helps you learn from mistakes. Rather than designing a structure that would take a year or more to analyze, you design it so that it can be assayed for given function or structure in a matter of days."

Many groups are pursuing protein design today, including academic researchers like Jane and Dave Richardson at Duke University, Bruce Erickson at the University of North Carolina, and Tom Blundell, Robin Leatherbarrow, and Alan Fersht in Britain. The successes have started to roll in. Japan, however, is unique in having an organization devoted exclusively to such projects: the Protein Engineering Research Institute (PERI) in Osaka. In 1990, PERI announced the successful design and construction of a de novo protein several times larger than any built before.

Speciality of Proteins

The main advantage of proteins is that they are familiar: a lot is known about them, and many tools exist for working with them. Yet proteins have disadvantages as well. Just because this design work is starting with proteins-soft, squishy molecules that are only marginally suitable for nanotechnology-doesn't mean it will stay within those limits.

De Grado points out "The fundamental goal of our work in de novo design is to be able to take the next step and get entirely away from protein systems." An early example is the work of Wallace Carothers of Du Pont, who used a de novo approach to studying the nature of proteins: Rather than trying to cut up proteins, he tried to build up things starting with amino acids and other similar monomers. In 1935, he succeeded in making nylon.

DeGrado explains "There is a deep philosophical belief at Du Pont in the ability of people to make molecules de novo that will do useful things. And there is a fair degree of commitment from the management that following that path will lead to products: not directly, and not always predictably, but they know that they need to support the basic science.

"I think ultimately we have a better chance at doing some really exciting things by de novo design, because our repertory should be much greater than that of nature. Think about the ability to fly: One could breed better carrier pigeons or one could design airplanes." The biology community, however, leans more toward ornithology than toward aerospace engineering. DeGrado's experience is that "a lot of biologists feel that if you aren't working with the real thing [natural proteins], you aren't studying biology, so they don't totally accept what we're doing. On the other hand, they recognize it as good chemistry."

Like the IBM physicists, protein designers are moved by a vision of molecular engineering. In 1989, Bill DeGrado predicted, "I think we'll be able to make catalysts or enzymelike molecules, possibly ones that catalyze reactions not catalyzed in nature." Catalysts are molecular machines that speed up chemical reactions: they form a shape for the two reacting molecules to fit into and thereby help the reaction move faster, up to a million reactions per second. New ones, for reactions that now go slowly, will give enormous cost savings to the chemical industry.

This prediction was borne out just a few months later, when Denver researchers John Stewart, Karl Hahn, and Wieslaw Klis announced their new enzyme, designed from scratch over a period of two years and built successfully on the first try. It's a catalyst, making some reactions go about 100,000 times faster. Nobel Prize-winning biochemist Bruce Merrifield believes that "if others can reproduce and expand on this work, it will be one of the most important achievements in biology or chemistry."

DeGrado also has longer term plans for protein design, beyond making new catalysts: "It will allow us to think about designing molecular devices in the next five to ten years. It should be possible ultimately to specify a particular design and build it. Then you'll have, say, proteinlike molecules that self-assemble into complex molecular objects, which can serve as machinery. But there's a limit to how small you can make devices. You'll shrink things down so far and then you won't be able to go any further, because you've reached molecular dimensions."

Mark Pearson shows that management at Du Pont also has this vision. Regarding the prospects for nanotechnology and assemblers, he remarked, "You know, it'll take money and effort and good ideas for sure. But to my way of thinking, there is no absolute fundamental limitation to preclude us from doing this kind of thing." He didn't say his company plans to develop nanotechnology, but such plans aren't really necessary. Du Pont is already on the nanotechnology path, for other-shorter-term, commercial-reasons. Like IBM, if they do decide to move quickly, they have the resources and forward-looking people needed to succeed.

Chemists, most of whom do not work on proteins, are the traditional experts in building molecular objects. As a group they've been building molecules for over a century, with ever increasing ability and confidence. Their methods are all indirect: They work with billions of atoms at a time-massive parallelism-but without control of the positions of their workpieces. The molecules typically tumble randomly in a liquid or gas, like pieces of a puzzle that may or may not fit together correctly when shaken together in a box. With clever design and planning, most pieces will join properly.

Chemists mix molecules on a huge scale (in our simulation view, a test tube holds a churning molecular swarm with the volume of an inland sea), yet they still achieve precise molecular transformations. Given that they work so indirectly, their achievements are astounding. This is, in part, the result of the enormous amount of work poured into the field for many decades. Thousands of chemists are working on molecular construction in the United States alone; add to that the chemists in Europe, in Japan, and in the rest of the world, and you have a huge community of researchers making great strides. Though it publishes only a one-paragraph summary of each research report, a guide to the chemical literature-Chemical Abstracts-covers several library walls and grows by many feet of shelf space every year.

Mixing Chemicals

An engineer would say that chemists (at least those specializing in synthesis) are doing construction work, and would be amazed that they can accomplish anything without being able to grab parts and put them in place. Chemists, in effect, work with their hands tied behind their backs. Molecular manufacturing can be termed "positional chemistry" or "positional synthesis," and will give chemists the ability to put molecules where they want them in three-dimensional space. Rather than trying to design puzzle pieces that will stick together properly by themselves when shaken together in a box, chemists will then be able to treat molecules more like bricks to be stacked. The basic principles of chemistry will be the same, but strategies for construction will become far simpler.

Without positional control, chemists face a problem something like this: Picture a giant glass barrel full of tiny battery-powered drills, buzzing away in all directions, vibrating around in the barrel. Your goal is to take a piece of wood and put a hole in just one specific spot. If you simply throw it in the barrel, it will be drilled haphazardly in many places.

To control the process, you must protect all the places you don't want drilled-perhaps by gluing protective pieces of metal over most of the wood surface. This problem-how to protect one part of a molecule while altering another part-has forced chemists to develop ever-cleverer ploys to build larger and larger molecules.

Chemists can achieve great things, but have focused much of their effort on duplicating molecules found in nature and then making minor variants.

As an example, take palytoxin, a molecule found in a Hawaiian coral. It was so difficult to make in the lab that it has been called "the Mount Everest of synthetic chemistry," and its synthesis was hailed as a triumph.

Other efforts are poured into making small molecules with unusual bonding, or molecules of remarkable symmetry, like "cubane" and "dodecahedrane" (shaped like the Platonic solids they are named after).

Chemists, at least in the United States, regard themselves as natural scientists even when their life's work is the construction of molecules by artificial means. Ordinarily, people who build things are called engineers.

And indeed, at the University of Tokyo the Department of Synthetic Chemistry is part of the Faculty of Engineering; its chemists are designing molecular switches for storing computer data. Engineering achievements will require work directed at engineering goals.

BUILDING MOLECULAR MACHINES

Molecular engineers working toward nanotechnology need a set of molecular building blocks for making large, complex structures. Systematic building-block construction was pioneered by Bruce Merrifield, winner of the 1984 Nobel Prize in Chemistry. His approach, known as "solid phase synthesis," or simply "the Merrifield method," is used to synthesize the long chains of amino acids that form proteins. In the Merrifield method, cycles of chemical reaction each add one molecular building block to the end of a chain anchored to a solid support.

This happens in parallel to each of trillions of identical chains, building up trillions of molecular objects with a particular sequence of building blocks. Chemists routinely use the Merrifield method to make molecules larger than palytoxin, and related techniques are used for making DNA in so-called gene machines: an ad from an Alabama company reads, "Custom DNA-Purified and Delivered in 48 hours." While it's hard to predict how a natural protein chain will fold-they weren't designed to fold predictably-chemists could make building blocks that are larger, more diverse, and more inclined to fold up in a single, obvious, stable pattern. With a set of building blocks like these, and the Merrifield method to string them together, molecular engineers could design and build molecular machines with greater ease.

To make a new molecule, both its structure and the procedure to make it must be designed. Compared to gigantic science projects like the Superconducting Supercollider and the Hubble Space Telescope, working with molecules can be done on a shoestring budget. Still, the costs of trying many different procedures add up. To help predict in advance what will work and what won't, designers turn to models.

You may have played with molecular models in chemistry class: colored plastic balls and sticks that fit together like Tinker Toys. Each color represents a different kind of atom: carbon, hydrogen, and so on. Even simple plastic models can give you a feel for how many bonds each kind of atom makes, how long the bonds are, and at what angles they are made. A more sophisticated form of model uses only spheres and partial spheres, without sticks. These colorful, bumpy shapes are called CPK models, and are widely used by professional chemists. Nobel laureate Donald Cram remarks that "We have spent hundreds of hours building CPK models of potential complexes and grading them for desirability as research targets." His research, like that of fellow Nobelists Charles J. Pedersen and Jean-Marie Lehn, has focused on designing and making medium-sized molecules that self assemble.

Although physical models can't give a good description of how molecules bend and move, computer-based molecules can. Computer-based modeling is already playing a key role in molecular engineering. As John Walker (a founder and leader of Autodesk) has remarked, "Unlike all of the industrial revolutions that preceded it, molecular engineering requires, as an essential component, the ability to design, model, and simulate molecular structures using computers." This has not gone unnoticed in the business community. John Walker's remark was part of a talk on nanotechnology given at Autodesk, a leader in computer-aided design and one of the five largest software firms in the United States. Soon after this talk, the company made its first major investment in the computer-aided design of molecules.

Manufacturers and architects know that designs for new products and buildings are best done on a computer, by computer-aided design (CAD). The new molecular design software can be called molecular CAD, and in its forefront are researchers such as Jay Ponder of the Yale University Department of Molecular Biophysics and Biochemistry. Ponder explains that "There's a strong link between what molecular designers are doing and what architects do. Michael Ward of Du Pont is designing a set of building blocks for a Tinker Toy set so that you can build larger structures. That's exactly what we're doing with molecular modeling techniques.

Getting to nanotechnology will require the work of experts in differing fields: chemists, who are learning how to make molecular machines; computer scientists, who are building the needed design tools; and perhaps STM and AFM experts, who can provide tools for molecular positioning. To make progress, however, these experts must do more than just work, they must work together.

Because nanotechnology is inherently interdisciplinary, countries that draw hard lines between their academic disciplines, as the United States does, will find that their researchers have difficulty communicating and cooperating. In chemistry today, a half-dozen researchers aided by a few tens of students and technicians is considered a large team. In aerospace engineering, enormous tasks like reaching the Moon or building a new airliner are broken down into tasks that are within the reach of small teams. All these small teams work together, forming a large team that may consist of thousands of engineers aided by many thousands of technicians. If chemistry is to move in the direction of molecular systems engineering, chemists will need to take at least a few steps in this direction.

In engineering, everyone knows that designing a rocket will require skills from many disciplines. Some engineers know structures, others know pumps, combustion, electronics, software, aerodynamics, control theory, and so on and so forth down a long list of disciplines. Engineering managers know how to bring different disciplines together to build systems.

In academic science, interdisciplinary work is productive and praised, but is relatively rare. Scientists don't need to cooperate to have their results fit together: they are all describing different parts of the same thing-nature-so in the long run, their results tend to come together into a single picture. Engineering, however, is different. Because it is more creative (it actually creates complex things), it demands more attention to teamwork. If the finished parts are going to work together, they must be developed by groups that share a common picture of what each part must accomplish. Engineers in different disciplines are forced to communicate; the challenge of management and team-building is to make that communication happen. This will apply to engineering molecular systems as much as it does to engineering computers, cars, aircraft, or factories.

Jay Ponder suggests that it's a question of perspective. "It's all a matter of what's perceived to be important by the different groups that have to come together to make this work: the chemists doing their bit and the computational people doing their bit. People have to come together and see the big picture. There are people who try to bridge the gaps, but they are rare compared to the people who just work in their own specialty." Progress toward nanotechnology will continue, and as it does, researchers trained as chemists, physicists, and the like will learn to talk to one another to solve new problems. They will either learn to think like engineers and work in teams, or they will be eclipsed by colleagues who do.

With all these problems, the advance toward nanotechnology steadily continues. Industry must gain ever-better control of matter to stay competitive in the world marketplace. The STM, protein engineering, and much of chemistry are driven by commercial imperatives. Focused efforts would yield faster advances, yet even without a clear focus, advances in this direction have an air of inevitability. As Bill DeGrado observes, "We really do have the tools. Experience has shown that when you have the analytic and synthetic tools to do things, in the end science goes ahead and does them-because they are doable."

Jay Ponder agrees: "Over the next few years, you're going to see slow evolutionary advances coming from people tinkering with molecular structures and figuring out their principles. People are going to work on a particular problem because they see some application for it or because they got grant funding for it. And in the process of doing something like improving a laundry detergent's ability to clean protein stains, Proctor and Gamble is going to help work out the principles for how to increase molecular stability, and to design spaces inside the molecules."

For a variety of reasons, Japan's contribution to nanotechnology research promises to be excellent. While the United States has generally pursued researching this area with little sense of long-term direction, it appears that Japan has begun to take a more focused approach. Researchers there already have clear ideas about molecular machines-about what might work and what probably won't. Japanese researchers are accustomed to a higher level of interdisciplinary contact and engineering emphasis than are Americans. In the United States, we prize "basic science," often calling it "pure science," as if to imply that practical applications are a form of impurity. Japan instead emphasizes "basic technology."

Nanotechnology is a basic technology, and the Japanese recognize it as such. Recent changes at the Tokyo Institute of Technology-Japan's equivalent of MIT-reflect their views of promising directions for future research. For many decades, Tokyo Tech has had two major divisions: a Faculty of Science and a Faculty of Engineering. To these is now being added a Faculty of Bioscience and Biotechnology, to consist of four departments: a Department of Bioscience, a Department of Bioengineering, a Department of Biomolecular Engineering, and what is termed a "Department of Biostructure." The creation of a new faculty in a major Japanese university is a rare event. What U.S. university has a department explicitly devoted to molecular engineering? Japan has both the departments at Tokyo Tech and Kyoto University's recently established Department of Molecular Engineering.

Japan's Institute for Physical and Chemical Research (RIKEN) has broad-based interdisciplinary strength. Hiroyuki Sasabe, head of the Frontier Materials Research Program at RIKEN, notes that the institute has expertise in organic synthesis, protein engineering, and STM technology. Sasabe says that his laboratory may need a molecular manipulator of the sort described in the next chapter to accomplish its goals in molecular engineering. Research consortia in Japan are also moving toward nanotechnology. The Exploratory Research for Advanced Technology Organization (ERATO) sponsors many three-to-five year projects in parallel, each with a specific goal. Consider the work in progress:

- o Yoshida Nanomechanism Project
- o Hotani Molecular Dynamic Assembly Project

- o Kunitake Molecular Architecture Project
- o Nagayama Protein Array Project
- o Aono Atomcraft Project

These focus on different aspects of gaining control over matter at the atomic level. The Nagayama Protein Array Project aims to use proteins as engineering materials to move toward making new molecular devices. The Aono Atomcraft Project does not involve nuclear power-as its translation might imply-but is instead an interdisciplinary effort to use an STM to arrange matter on the atomic scale.

At some point, work on nanotechnology must move beyond spin-offs from other fields and undertake the design and construction of molecular machinery. This shift from opportunistic science to organized engineering requires a change in attitude. In this, Japan leads the United States.

Molecular nanotechnology will emerge step by step. Major milestones, such as the engineering of proteins and the positioning of individual atoms, have already been passed. To get a sense of the likely pace of developments, we need to look at how various trends fit together.

Computer-based molecular-modeling tools are spawning computer-aided design tools. These will grow more capable. The underlying technology base-computer hardware-has for decades been improving in price and performance on a steeply rising curve, which is generally expected to continue for many years. These advances are quite independent of progress in molecular engineering, but they make molecular engineering easier, speeding advances. Computer models of molecular machines are beginning to appear, and these will whet the appetites of researchers.

Progress in engineering molecular machines, whether using proximal probes or self-assembly, will eventually achieve striking successes; the objectives of research in Japan will begin to draw serious attention; understanding of the long-term promise of molecular engineering will become more widespread.

Some combination of these developments will eventually lead to a serious, public appraisal of what these technologies can achieve-and then the world of opinion, funding, and research fashion will change. Before, advances will be steady but haphazard; afterward, advances will be driven with the energy that flows into major commercial, military, and medical research programs, because nanotechnology will be recognized as furthering major commercial, military, and medical goals. The timing of subsequent events depends largely on when this threshold of serious attention is reached.

In making time estimates, people are prone to assume that a large change must take a long time. Most do, but not all. Pocket calculators had a dramatic effect on the slide-rule industry: they replaced it. The speed of this change caught the slide rule moguls by surprise, but the pace of progress in electronics didn't slow down merely to suit their expectations. One can argue that nanotechnology will be developed fast: many countries and companies will be competing to get there first. They will be driven onward both by the immense expected benefits-in many areas, including medicine and the environment-as well as by potential military applications. That is a powerful combination of motives, and competition is a powerful accelerator.

A counterargument, though, suggests that development will be slow: anyone who has done anything of significance in the real world of technology-doing a scientific experiment, writing a computer program, bringing a new product to market-knows that these goals take longer than expected. Indeed, Hofstadter's Law states that projects take longer than expected, even when Hofstadter's Law is taken into account. This principle is a good guide for the short term, and for a single project.

The situation differs, though, when many different approaches are being explored by many different groups over a period of years. Most projects may take longer than expected, but with many teams trying many approaches, one approach may prove faster than expected. The winner of a race is always faster than the average runner. John Walker notes, "The remarkable thing about molecular engineering is that it looks like there are many different ways to get there and, at the moment, rapid progress is being made along every path-all at the same time."

Also, technology development is like a race run over an unmapped course. When the first runners reach the top of a hill, they may see a shortcut. A trailing runner may decide to crash off into the bushes, and stumble across a bicycle and a paved road. The progress of technology is seldom predictable because progress often reveals new directions. How close we are to goal depends on whether technological advances are a constant pace of accelerating. In this diagram, the dashed line represents the current level of technology, and the large dot in the upper right represents a goal such as nanotechnology. With a straight-line advance, it's easier to estimate how far away a goal is. With an accelerating advance, a goal can be reached with little warning.

So how can we estimate a date for the arrival of nanotechnology? It's safest to take a cautious approach: When anticipating benefits, assume it's far off; when preparing for potential problems, assume it's right around the corner. The old folk saying applies: Hope for the best, prepare for the worst. Any dates assigned to "far off" and "right around the corner" can be no better than educated guesses-molecular behavior can be calculated, but not technology timetables of this sort. With those caveats, we would estimate that general-purpose molecular assemblers will likely be developed in the early decades of the twenty-first century, perhaps in the first.

John Walker, whose technological foresight has led Autodesk from start-up to a dominant role in its industry, points out that not long ago, "Many visionaries intimately familiar with the developments of silicon technology still forecast it would take between twenty and fifty years before molecular engineering became a reality. This is well beyond the planning horizon of most companies. But recently, everything has begun to change." Based on the new developments, Walker places his bet: "Current progress suggests the revolution may happen within this decade, perhaps starting within five years.

13

Threshold of Nanotechnological

In this chapter, we outline how emerging technologies can lead to nanotechnology. The actual path to nanotechnology-the one that history books will record-could emerge from any one of the research directions in physics, biochemistry, and chemistry recounted in the last chapter, or (more likely) from a combination of them. The availability of so many good options builds confidence that the goal can be reached, even while it decreases confidence that some particular path will be fastest. To see how advances might cross the gap from present technology to early nanotechnology, let's follow one path out of the many possible.

One way to bridge the gap would through the development of an AFM-based molecular manipulator capable of doing primitive molecular manufacturing. This device would combine a simple molecular device-a molecular gripper-with an AFM positioning mechanism. An AFM can move its tip with precision; a molecular manipulator would add a gripper to the tip to hold a molecular tool. A molecular manipulator of this kind would guide chemical reactions by positioning molecules, like a slow, simple, but enormous assembler. (In our standard simulation view, where a molecular assembler arm fits in a room, the AFM apparatus of a molecular manipulator would be the size of a moon.) Despite its limits, an AFM molecular manipulator will be a striking advance. How might this advance occur? Since we're choosing one path out of many possible, we may as well include more details and tell a story.

DEVELOPING A MOLECULAR MANIPULATOR

Several years ago, researchers at the University of Brobdingnag began work on developing a molecular manipulator. To reach this goal, a team of a dozen physicists, chemists, and protein researchers banded together (some working full time, some part time) and began the creative teamwork needed to solve the basic problems.

First they needed to attach a gripper to an AFM tip. As grippers, they chose fragments of antibody molecules, the selectively sticky proteins that the immune system uses to bind and identify germs. If they could get the "back" of the molecule stuck onto a tip, then the "front" could bind and hold molecular tools. (The advantage of antibody fragments was this: freedom of tool choice. Since the late 1980s, researchers had been able to generate antibodies able to bind almost any preselected molecule-or molecular tool.)

A molecular manipulator would bind and position reactive molecular tools to build up a workpiece, molecule by molecule. In parallel, the U. Brob AFM researchers worked on placing tips in a precise location and then holding them there with atomic accuracy for seconds at a time. This proved straightforward. They used techniques developed elsewhere during the early 1990s, adding only modest refinements.

They now had their gripper and a way of putting it where they wanted it, but they needed a set of tools. The gripper was like the chuck of a drill, waiting to have different bits fitted into its tool-holder slot. So as the final step, the synthetic chemists on the team made a dozen different molecular tools, all identical at one end but different at the other. The similar parts all bound to the same antibody tool-holder, slotting neatly into position. The different parts were all chemically reactive in different ways.

Developing the molecular tool kit was the toughest part of the project; it took about as much work as had gone into duplicating the palytoxin molecule back in the 1980s. None of the tasks in the project demanded the solution of a deep scientific puzzle, and none demanded the solution of a notoriously difficult engineering problem. Each task had many possible solutions, the problem was to find a compatible set of solutions and apply them. After a few years, the solutions came together and the U. Brob research team began building new molecules by molecular manipulation. Now many teams are doing likewise.

GRIPPERS AND TOOLS

To build something with the U. Brob team's AFM-based molecular manipulator system, you use it as follows: First, choose a surface to build on and place it under the tip in a pool of liquid. Then dunk the AFM tip into the liquid, bringing it down to the surface, and back it off a little. Construction can now begin as soon as a tool is loaded into the gripper.

Tubes and pumps can flow different liquids over the surface and past the gripper, carrying different tool molecules. If you want to do something with a tool of Type A, you wash in the proper liquid. Once it is in the gripper, you can use the AFM mechanism to move it around and put it where you want it. Move it up to the surface at a convenient spot, wait a few seconds, and it reacts, forming a bond and leaving a molecular fragment attached to the spot you chose. To add a different fragment, you can use a tool of Type B: you back up the tip, flow in a fresh liquid carrying the new tools, and in a moment a tool of the new type is bound in place and ready to apply, either on or alongside the first spot. Step by step, you build up a precise molecular structure.

Each step takes only seconds. Molecular tools pop into the gripper in a fraction of a second, and used tools pop off at the same rate. Once the tip has positioned a molecule, it reacts quickly, about a million times faster than unwanted reactions at other sites. In this way, the molecular manipulator gives good control of where reactions will occur (though it is not as reliable as an advanced assembler would be). It is fairly fast by a chemist's standards-per cycle-but still a million times slower than an advanced assembler. It can perform a variety of steps, but isn't as flexible and capable as an advanced assembler. In short, it is hardly the last word in nanotechnology, yet is a great advance over what has gone before.

PRODUCTS

With its ability to accelerate desired reactions by a factor of a million or so, the U. Brob team's molecular manipulator can perform 10,000 to 100,000 steps with good reliability. Back in the 1980s, chemists making protein molecules struggled to perform just one hundred steps. The U. Brob research team (and its many imitators) can now build structures that are stronger and easier to design than proteins: not floppy, folded chains, but rugged objects held together by a sturdy network of bonds. Though not as strong and dense as diamond, these structures are like bits of a tough engineering plastic. A specially adapted computer-aided design system makes it easy to design molecular objects made from these materials. Yet the AFM-based molecular manipulator has one grave disadvantage: It does chemistry one molecule at a time, and it ties up a machine as expensive as a car for hours or days to produce that one large molecule. Some molecules, though, are valuable enough to be worth building even one at a time. These draw prompt attention.

A single molecule isn't much use as a dye, a drug, or a floor wax, but it can have substantial value if it provides useful information. The U. Brob team quickly publishes a pile of scientific papers based on experiments with single molecules: they build a molecule, probe it, report the results, and build another. Some of these results show chemists elsewhere in the multibillion-dollar chemical industry how to design new catalysts, molecules that can help make other molecules more cheaply, cleanly, and efficiently. This information is worth a lot.

Three new products of special interest are among the first to be made. The first-molecular electronics-begins with experiments conducted by a research group at a computer chip company. They use their molecular manipulator to build single molecules and probe them, gradually learning how to build the parts needed for molecular electronic computers. These new computers don't immediately become practical, because the costs are too high for making such large molecules with AFM-based technology. Yet some companies begin to produce simpler molecular electronic devices for use in sensors and specialized high-speed signal processing. A specialty industry is born and begins to expand.

The second product is a gene reader, a complex molecular device built on the surface of a chip. The biologists who built the reader combined proteins borrowed from cells with special-purpose molecular machines designed from scratch. The result was a molecular system that binds DNA molecules and pulls them past a read-head-like tape through a tape recorder. The device works as fast as some naturally occurring molecular machines that read DNA, with one key advantage: it outputs its data electronically. At that speed, a single device can read a human genome in about a year. Though still too expensive for a doctor's office, these readers are promptly in great demand from research laboratories. Another small industry is born.

The third product is far more important, in the long run: replacement tips for molecular manipulators, grippers, and tools that are better than the originals. With these new, more versatile devices, researchers are now building more ambitious products and tools.

CHEMIST'S TOOLS

How did the chemists achieve this? During the years when the U. Brob team was developing the molecular manipulator, researchers working in protein science and synthetic chemistry had made better and better systems of molecular building blocks. Chemists were well prepared for doing this: by the late 1980s, it had become possible to build stable structures the size of medium-sized protein molecules, and work had begun to focus on making these molecules perform useful work by binding and modifying other molecules. Chemists learned to use these sophisticated catalysts-early molecular devices-to make their own work easier by helping in the manufacture of still more large molecules.

Another traditional chemist's tool was software for doing computer-aided design. The early software designed by Jay Ponder and Frederic Richards of Yale University ultimately led to semi-automatic tools for designing molecules of a particular shape and function. Chemists then could easily design molecules that would self-assemble into larger structures, several tens of nanometers across.

Molecular Construction Machines

These advances in software and chemical synthesis let the U. Lill team tackle the task of building a primitive version of a molecular assembler. Although they couldn't build anything as complex as a nanocomputer or as stiff as diamond, they didn't need to. Their design used sliding molecular rods to position a molecular gripper much like the gripper used at U. Brob, again using the surrounding liquid to control which tool the gripper held. Instead of an AFM's electronic controls, they used the surrounding liquid to control the position of the rods as well. In a neutral solution, the rods would withdraw; in an acid solution, they would extend. How far they moved depended on what other molecules were around to lodge in special pockets and block the motion.

Their primitive assemblers built much the same sorts of products that the U. Brob molecular manipulator did; the tools were similar, and speed and accuracy were about the same. Yet there was one dramatic advantage: About 1,000,000,000,000,000,000,000 U. Lill assemblers could fit in the space occupied by one U. Brob manipulator, and it was easy to produce a mere 1,000,000,000,000,000 times as much product at the same cost. With the first, primitive assemblers, construction was slow because each step required new liquid baths and several seconds of soaking and waiting, and a typical product might take thousands of steps. Nonetheless, the U. Lill team made a lot of money licensing their technology to researchers trying to commercialize products they had first researched with the U. Brob machine. After starting an independent company (Nanofabricators, Inc.), they poured their research efforts into building better machines.

Within a few years, they had assemblers with multiple grippers, each loaded with a different kind of tool; flashes of colored light would flip molecules from state to state (they copied these molecules from the pigments of the retina of the eye); flipping molecules would change tools and change rod positions. Soaking and waiting become a thing of the past, and soon they were pouring out parts that, when mixed with liquid and added to dishes with special blank chips would build up the dense memory layers that made possible the Pocket Library. That was when things started moving fast. The semiconductor industry went the way of the vacuum tube industry. Money and talent poured into the new technology. Molecular CAD tools got better, assemblers made it easy to build what was designed, and fast production and testing made molecular engineering as easy as playing with software. Assemblers got better, faster, and cheaper. Researchers used assemblers to build nanocomputers, and nanocomputers to control better, faster assemblers. Using tools to build better tools is an ancient story. Within a decade, almost anything could be made by molecular manufacturing, and was.

MINI COMPUTERS

While the first, primitive assemblers were controlled by changing what molecules are in the solution around the device, getting the speed and accuracy wanted for large-scale manufacturing takes real computation. Carl's setup uses a combination of special-purpose molecule processors and general-purpose assemblers, all controlled and orchestrated by nanocomputers.

Computers back in the 1990s used microelectronics. They worked by moving electrical charge back and forth through conducting paths-wires, in effect-using it to block and unblock the flow of charge in other paths. With nanotechnology, computers are built from molecular electronics. Like the computers of the 1990s, they use electronic signals to weave the patterns of digital logic. Being made of molecular components, though, they are built on a much smaller scale than 1990s computers, and work much faster and more efficiently. On the scale of our simulated molecular world, 1990s computer chips are like landscapes, while nanocomputers are like individual buildings. Carl's desktop PC contains over a trillion nanocomputers, enough to out-compute all the microelectronic computers of the twentieth century put together.

Back in the dark ages of the 1980s, an exploratory engineer proposed that nanocomputers could be mechanical, using sliding rods instead of moving electrons as shown in Figure 8. These molecular mechanical computers were much easier to design than molecular electronic computers would have been. They were a big help in getting some idea of what nanotechnology could do.

An electronic transistor lets current flow when a negative electric charge is applied and blocks current when a positive charge is applied. The mechanical "transistor" lets the horizontal rod move when the vertical rod is down, and blocks the horizontal rod when the vertical rod is up. Either device can be used to build logic gates and computers.

Even back then, it was pretty obvious that mechanical computers would be slower than electronic computers. Carl's molecular electronic PC would have been no great surprise, though nobody knew just how to design one. When nanotechnology actually arrived and people started competing to build the best possible computers, molecular electronics won the technology race. Still, mechanical nanocomputers could have done all the nanocomputing jobs at Desert Rose: ordinary, everyday molecular manufacturing just doesn't demand the last word in computer performance.

For Carl, the millions of nanocomputers in the milky waters of his building ponds are just extensions of machines on his desk, machines there to help him run his business and deliver products to his customers-or, in the case of the Red Cross emergency, to help provide time-critical emergency supplies. By reserving those three separate ponds, Carl can either build three different kinds of equipment for the Red Cross or use all the ponds to mass-produce the first thing on the Red Cross list: emergency shelters for ten thousand people. The software is ready, the plumbing is fine, the drums of building materials are all topped up, the Special Mix for this job is loaded: the build is ready to start. "Okay," Carl tells the computer, "build Red Cross tents." Computer talks to nanocomputers. In all three pools, nanocomputers talk to assemblers. The build begins.

Assembling

Some of the building done at Desert Rose Industries uses assemblers much like the ones we saw in the first hall of the plant tour, back in the simulated molecular world of the Silicon Valley Faire. As seen in simulation, they are big, slow, computer-controlled things moving molecular tools. With the right instructions and machinery to keep them supplied with molecular tools, these general-purpose assemblers can build almost anything. They're slow, though, and take a lot of energy to run. Some of the building uses special-purpose assembly systems in the molecule-processing style, like the systems in the basement we saw in the tour of a simulated molecular factory. The special-purpose systems are all moving belts and rollers, but no arms. This is faster and more efficient, but for quantity orders, cooling requirements limit the speed.

It's faster to use larger, prefabricated building blocks. Desert Rose uses these for most of their work, and especially for rush orders like the one Carl just set up. Their underground warehouse has room-sized bins containing upward of a thousand tons of the most popular building blocks, things like structural fibers. They're made at plants on the West Coast and shipped here by subway for ready use. Other kinds are made on site using the special-purpose assemblers. Carl's main room has several cabinet-sized boxes hooked up to the plumbing, each taking in raw materials, running them through this sort of specialized molecular machinery, and pumping out a milky syrup of product. One syrup contains motors, another one contains computers, and another is full of microscopic plug-in light sources. All go into tanks for later use.

Now they're being used. The mix for the Red Cross tent job is mostly structural fiber stronger than the old bulletproof-vest materials. Other building blocks also go in, including motors, computers, and dozens of little struts, angle brackets, and doohickies. The mix would look like someone had stirred together the parts from a dozen toy sets, if the parts were big enough to see. In fact, though, the largest parts would be no more than blurry dots, if you saw one under a normal optical microscope.

The mix also contains block-assemblers, floating free like everything else. These machines are big, about like an office building in our simulation view with the standard settings. Each has several jointed arms, a computer, and several plugs and sockets. These do the actual construction work.

To begin the build, pumps pour the mix into a manufacturing pond. The constant tumbling motions of microscopic things in liquids would be too disorganized for building anything so large as a tent, so the block-assemblers start grabbing their neigbours. Within moments, they have linked up to form a framework spread through the liquid. Now that they are plugged together, they divide up jobs, and get to work. Instructions pour in from Carl's desktop computer.

The block-assemblers use sticky grippers to pull specific kinds of building blocks out of the liquid. They use their arms to plug them together. For a permanent job, they would be using blocks that bond together chemically and permanently. For these temporary tents, though, the Red Cross design uses a set of standard blocks that are put together with amazingly ordinary fasteners: these blocks have snaps, plugs, and screws, though of course the parts are atomically perfect and the threads on the screws are single helical rows of atoms. The resulting joints weaken the tent's structure somewhat, but who cares? The basic materials are almost a hundred times stronger than steel, so there is strength to waste if it makes manufacturing more convenient.

Fiber segments snap together to make fabrics. Some segments contain motors and computers, linked by fibers that contain power and data cables. Struts snap together with more motors and computers to make the tent's main structures. Special surfaces are made of special building blocks. From the human perspective, each tent is a lightweight structure that contains most of the conveniences and comforts of an apartment: cooking facilities, a bathroom, beds, windows, air conditioning, specially modified to meet the environmental demands of the quake-stricken country. From a builder's perspective, especially from a nanomachine's point of view, the tent is just structure slapped together from a few hundred kinds of prefab parts. In a matter of seconds, each block-assembler has put together a few thousand parts, and its section of the tent is done. In fact, the whole thing is done: many trillions of hands make light work. A crane swings out over the pond and starts plucking out tent packages as fresh mix flows in.

Maria's concern has drawn her back to the plant to see how the build is going. "It's coming along," Carl reassures her. "Look, the first batch of tents is out." In the warehouse, the first pallet is already stacked with five layers of dove-gray "suitcases": tents dried and packed for transport. Carl grabs a tent by the handle and lugs it out the door. He pushes a tab on the corner labeled "Open," and it takes over a minute to unfold to a structure a half-dozen paces on a side. The tent is big, and light enough to blow away if it didn't cling to the ground so tightly. Maria and Carl tour the tent, testing the appliances, checking the construction of furniture: everything is extremely lightweight compared to the bulk-manufactured goods of the 1990s, tough but almost hollow.

Like the other structures, the walls and floors are full of tiny motors and struts controlled by simple computers like the ones used in twentieth-century cars, televisions, and pinball machines. They can unfold and refold. They can also flex to produce sound like a high-quality speaker, or to absorb sound to silence outdoors racket. The whole three-room setup is small and efficient, looking like a cross between a boat cabin and a Japanese business hotel room. Outside, though, it is little more than a box.

BEHIND THE SCENES AND AFTERWARD

Desert Rose Industries and other manufacturers can make almost anything quickly and at low cost. That includes the tunneling machines and other equipment that made the subway system they use for shipping. Digging a tunnel from coast to coast now costs less than digging a single block under New York City used to. It wasn't expensive to get a deep-transit terminal installed in their basement. Just as the tents aren't mere bundles of canvas, these subways aren't slow things full of screeching, jolting metal boxes. They're magnetically levitated to reach aircraft speeds-as experimental Japanese trains were in the late 1980s-making it easy for Carl and Maria to give their customers quick service. There's still a road leading to the plant, but nobody's driven a truck over it for years.

They only take in materials that they will eventually ship out in products, so there's nothing left over, and no wastes to dump. One corner of the plant is full of recycling equipment. There are always some obsolete parts to get rid of, or things that have been damaged and need to be reworked. These get broken down into simpler molecules and put back together again to make new parts.

The gunk in the manufacturing ponds is water mixed with particles much finer than silt. The particles-fasteners, computers, and the rest-stay in suspension because they are wrapped in molecular jackets that keep them there. This uses the same principle as detergent molecules, which coat particles of oily dirt to float them away.

Though it wouldn't be nutritious or appetizing, you could drink the tent mix and be no worse for it. To your body, the parts and their jackets, and even the nanomachines, would be like so many bits of grit and sawdust. (Grandma would have called it roughage.)

Carl and Maria get their power from solar cells in the road, which is the only reason they bothered having it paved. In back of their plant stands what looks like a fat smokestack. All it produces, though, is an updraft of clean, warm air. The darkly paved road, baking in the New Mexico sun, is cooler than you might expect: it soaks up solar energy and makes electricity, instead of just heat. Once the power is used, it turns back into heat, which has to go somewhere. So the heat rises from their cooling tower instead of the road, and the energy does useful work on the way.

Some products, like rocket engines, are made more slowly and in a single piece. This makes them stronger and more permanent. The tents, though, don't need to be superstrong and are just for temporary use. A few days after the tents go up, the earthquake victims start to move out into new housing (permanent, better-looking, and very earthquake resistant). The tents get folded and shipped off for recycling.

Recycling things built this way is simple and efficient: nanomachines just unscrew and unsnap the connectors and sort the parts into bins again. The shipments Desert Rose gets are mostly recycled to begin with. There's no special labeling for recycled materials, because the molecular parts are the same either way.

For convenience (and to keep the plant small), Carl and Maria get most of their parts prefabricated, even though they can make almost anything. They can even make more production equipment. In one of their manufacturing ponds, they can put together a new cabinet full of special-purpose assemblers. They do this when they want to make a new type of part in-house. Like parts, the part-assemblers are made by special-purpose assemblers. Carl can even make big vats in medium-size vats, unfolding them like tents.

If Desert Rose Industries needed to double capacity, Carl and Maria could do it in just a few days. They did this once for a special order of stadium sections. Maria got Carl to recycle the new building before its shadow hurt their cactus garden.

In the Desert Rose Industries scenario, manufacturing has become cheap, fast, clean, and efficient. Using fast, precise machines to handle matter in molecular pieces makes it easy for nanotechnology to be fast, clean, and efficient. But for it to be cheap, the manufacturing equipment has to be cheap.

The Desert Rose scenario shows how this can work. Molecular-manufacturing equipment can be used to make all the parts needed to build more molecular manufacturing equipment. It can even build the machines needed to put the parts together. This resembles an idea developed by NASA for a self-expanding manufacturing complex on the Moon, but made faster and simpler using molecular machines and parts.

In the early days of nanotechnology, there won't be as many different kinds of machines as there are at Desert Rose. One way to build a lot of molecular manufacturing equipment in a reasonable time would be to make a machine that can be used to make a copy of itself, starting with special but simple chemicals. A machine able to do this is called a "replicator." With a replicator and a pot full of the right fuel and raw materials, you could start with one machine, then have two, four, eight, and so on.

This doubling process soon makes enough machines to be useful. The replicators-each including a computer to control it and a general-purpose assembler to build things-could then be used to make something else, like the tons of specialized machines needed to set up a Desert Rose manufacturing plant. At that point, the replicators could be discarded in favour of those more efficient machines.

Replicators are worth a closer look, though, because they show how quickly molecular manufacturing systems can be used to build more manufacturing equipment. If we were in one of our standard simulation views, the submicroscopic device at the top of the picture would be like a huge tank, three stories tall when lying on its side. Most of its interior is taken up by a tape memory system that tells how to move the arm to build all the parts of the replicator, except the tape itself. The tape gets made by a special tape-copying machine. At the right-hand end of the replicator are pores for bringing in fuel and raw-material molecules, and machinery for processing them. In the middle are computer-controlled arms, like the ones we saw on the plant trip. These do most of the actual construction.

The steps in the cycle-using a copy to block the tube, beginning a fresh copy, then releasing the old one-illustrate one way for a machine to build a copy of itself while floating in a liquid, yet doing all its construction work inside, in vacuum. (It's easier to design for vacuum, and this is exploratory-engineering work, so easier design is better design.) Calculations suggest that the whole construction cycle can be completed in less than a quarter hour, since the replicator contains about a billion atoms, and each arm can handle about a million atoms per second. At that rate, one device can double and double again to make trillions in about ten hours.

Each replicator just sits in a chemical bath, soaking up what it needs and making more replicators. Eventually, either the special chemicals run out or other chemicals are added to signal them to do something else. At that point, they can be reprogrammed to produce anything else you please, so long as it can be extruded from the front. The products can be long, and can unfold or be pieced together to make larger objects, so the size of these initial replicators-smaller than a bacterium-would be only a temporary limitation.

From the molecular manipulators and primitive assemblers described in the last chapter, the most likely path to nanotechnology leads to assemblers with more and more general capabilities. Still, efficiency favours special-purpose machines, and the Desert Rose scenario didn't make much use of general assemblers. Why bother making general-purpose assemblers in the first place?

To see the answer, turn the question around and ask, Why not build such a tool? There is nothing outstandingly difficult about a general assembler, as molecular machinery goes. It will just be a device with good, flexible positional control and a system to feed it a variety of molecular tools. This is a useful, basic capability. General-purpose assemblers could always be replaced by a lot of specialized devices, but to build those specialized devices in the first place, it makes sense to come up with a more flexible, general-purpose system that can just be reprogrammed.

So, general purpose machines are likely to find use in making short production runs of more specialized devices. Ralph Merkle, a computers and security expert at Xerox Palo Alto Research Center, sees this as paralleling the way manufacturing works today: "General purpose devices could do many tasks, but they'll do them inefficiently. For any given task, there will be one or a few best ways of doing it, and one or a few special purpose devices that are finely tuned to do that one task. Nails aren't made by a general-purpose machine shop, they're made by nail-making machines.

Making nails with a general-purpose machine shop would be more expensive, more difficult, and more time-consuming. Likewise, in the future we won't see a proliferation of general-purpose self-replicating systems, we'll see specialization for almost every task."We've surveyed a lot of devices: assemblers of various flavors, nanocomputers, disassemblers, replicators, and others. What's important about these is not the exact distinctions between them, but the capabilities that they will give and the effects they will have on human lives. Again, we are suspending discussion of potential misapplications until later.

If we tease apart the implications of what we've seen in the Desert Rose scenario, we can analyze some of the key impacts of molecular manufacturing in industry, science, and medicine.

14

Industry and Nanotechnology

At its base, nanotechnology is about molecular manufacturing, and manufacturing is the basis of much of today's industry. This is why Desert Rose made a good starting point for describing the possibilities of a nanotechnological world. From an industrial perspective, it makes sense to think of nanotechnology in terms of products and production.

NEW PRODUCTS

Today, we handle matter crudely, but nanotechnology will bring thorough control of the structure of matter, the ability to build objects to atom-by-atom specifications. This means being able to make almost anything. By comparison, even today's range of products will feel very limited. Nanotechnology will make possible a huge range of new products, a range we can't envision today. Still, to get a feel for what is possible, we can look at some easily imagined applications.

RELIABLE PRODUCTS

Today, products often fail, but for failures to occur-for a wing to fall off an airplane, or a bearing to wear out-a lot of atoms have to be out of place. In the future, we can do better. There are two basic reasons for this:

better materials and better quality control, both achieved by molecular manufacturing. By using materials tens of times stronger than steel, as Desert Rose did, it will be easy to make things that are very strong, with a huge safety margin. By building things with atom-by-atom control, flaws can be made very rare and extremely small-nonexistent, by present standards.

With nanotechnology, we can design in big safety margins and then manufacture the design with near-perfection. The result will be products that are tough and reliable. (There will still be room for bad designs, and for people who wish to take risks in machines that balance on the edge of disaster.)

INTELLIGENT PRODUCTS

Today, we make most things from big chunks of metal, wood, plastic, and the like, or from tangles of fibers. Objects made with molecular manufacturing can contain trillions of microscopic motors and computers, forming parts that work together to do something useful. A climber's rope can be made of fibers that slide around and reweave to eliminate frayed spots. Tents can be made of parts that slide and lock to turn a package into a building. Walls and furniture can be made to repair themselves, instead of passively deteriorating.

On a mundane level, this sort of flexibility will increase reliability and durability. Beyond this, it will make possible new products with abilities we never imagined we needed so badly. And beyond even this, it will open new possibilities for art.

INEXPENSIVE PRODUCTION

Today, production requires a lot of labor, either for making things or for building and maintaining machines that make things. Labor is expensive, and expensive machines make automation expensive, too. In the Desert Rose scenario, we got a glimpse of how molecular manufacturing can make production far less expensive than it is today. This is perhaps the most surprising conclusion about nanotechnology, so we'll take a closer look at it in the next chapter.

CLEAN PRODUCTION

Today, our manufacturing processes handle matter sloppily, producing pollution. One step puts stuff where it shouldn't be; the next washes it off the product and into the water supply. Our transportation system worsens the problem as unreliable trucks and tankers spill noxious chemicals over the land and sea. Everything is expensive, so companies skimp on even the half-effective pollution controls that we know how to build.

Nanotechnology will mean greater control of matter, making it easy to avoid pollution. This means that a little public pressure will go a long way toward a cleaner environment. Likewise, it will make it easy to increase efficiency and reduce resource requirements. Products, like the Red Cross tents at Desert Rose, can be made of snap-together, easily recyclable parts. Sophisticated products could even be made from biodegradable materials. Nanotechnology will make it easy to attack the causes of pollution at their technological root.

Nanotechnology will have great applications in the field of industry, much as transistors had great applications in the field of vacuum tube electronics, and democracy had great applications in the field of monarchy. It will not so much advance twentieth-century industry as replace it-not all at once, but during a thin slice of historical time.

Chemistry

Today, chemists work with huge number of molecules and study them using clever, indirect techniques. Making a new molecule can be a major project, and studying it can be another. Molecular manufacturing will help chemists make what they want to study, and it will help them make the tools they need to study it. Nanoinstruments will be used to prod, measure, and modify molecules in a host of ways, studying their structures, behaviors, and interactions.

Materials

Today, materials scientists make new superconductors, semiconductors, and structural materials by mixing and crushing and baking and freezing, and so forth. They dream of far more structures than they can make, and they stumble across more things than they plan. With molecular manufacturing, materials science can be much more systematic and thorough. New ideas can be tested because new materials can be built according to plan (rather than playing around, groping for a recipe).

This need not rule out unexpected discoveries, since experiments-even blind searches-will go much faster. A few tons of raw materials would be enough to make a billion samples, each a cubic micron in size. In all of history so far, materials scientists have never tested so many materials. With nanoinstruments and nanocomputers, they could. One laboratory could then do more than all of today's materials scientists put together.

Biology

Today, biologists use a host of molecular devices borrowed from biology to study biology. Many of these can be viewed as molecular machines. Nanotechnology will greatly advance biology by providing better molecular devices, better nanoinstruments. Some cells have already been mapped in amazing molecular detail, but biology still has far to go. With nanoinstruments (including molecule-by-molecule disassemblers), biologists will at last be able to map cells completely and study their interactions in detail. It will become easy not only to find molecules in cells, but to learn what they do. This will help in understanding disease and the molecular requirements for health, enormously advancing medicine.

Computation

Today, computers range from a million to a billion times faster than an old desktop adding machine, and the results have been revolutionary for science. Every year, more questions can be answered by calculations based on known principles of physics. The advent of nanocomputers-even slow, miserable, mechanical nanocomputers-will give us practical machines with a trillion times the power of today's computers (essentially by letting us package a trillion computers in a small space, without gobbling too much money or energy.) The consequences will again be revolutionary.

Physics

The known principles of physics are adequate for understanding molecules, materials, and cells, but not for understanding phenomena on a scale that would still be submicroscopic if atoms were the size of marbles. Nanotechnology can't help here directly, but it can provide manufacturing facilities that will make huge particle accelerators economical, where today they strain national budgets.

More generally, nanotechnology will help science wherever precision and fine details are important. Science frequently proceeds by trying small variations in almost identical experiments, comparing the results. This will be easier when molecular manufacturing can make two objects that are identical, molecule by molecule. In some areas, today's techniques are not only crude, but destructive. Archaeological sites are unique records of the human past, but today's techniques throw away most information during the dig, by accident. Future archaeologists, able to sift soil not speck by speck but molecule by molecule, will be grateful indeed to those archaeologists who today leave some ground undisturbed.

Of all the areas where the ability to manufacture new tools is important, medicine is perhaps the greatest. The human body is intricate, and that intricacy extends beyond the range of human vision, beyond microscopic imaging, down to the molecular scale. "Molecular medicine" is an increasingly popular term today, but medicine today has only the simplest of molecular tools. As biology uses nanoinstruments to learn about disease and health, we will learn the physical requirements for restoring and maintaining health. And with this knowledge will come the tools needed to satisfy those requirements-tools ranging from improved pharmaceuticals to devices able to repair cells and tissues through molecular surgery.

Advanced medicine will be among the most complex and difficult applications of nanotechnology. It will require great knowledge, but nanoinstruments will help gather this knowledge. It will pose great engineering challenges, but computers of trillionfold greater power will help meet those challenges. It will solve medical problems on which we spend billions of dollars today, in hopes of modest improvements.

Today, modern medicine often means an expensive way to prolong misery. Will nanomedicine be more of the same? Any reader over the age of, say, thirty knows how things start to go wrong: an ache here, a wrinkle there, the loss of an ability. Over the decades, the physical quality of life declines faster and faster-the limits of what the body can do become stricter-until the limits are those of a hospital bed. The healing abilities we have when young seem to fade away. Modern medical practice expends the bulk of its effort on such things as intensive care units, dragging out the last few years of life without restoring health.

Truly advanced medicine will be able to restore and supplement the youthful ability to heal. Its cost will depend on the cost of producing things more intricate than any we have seen before, the cost of producing computers, sensors, and the like by the trillions. To understand the prospects for medicine, like those for science and industry, we need to take a closer look at the cost of molecular manufacturing.

In earlier chapters, we have stepped forward and backward through time. The last step was a big one, leaping from small laboratory devices to the high-capacity industrial facility of the Desert Rose scenario. Our narrative crossed this gap in a single leap, but the world won't. To understand how nanotechnology might unfold, it makes sense to look at some of its easier and more difficult applications. The result won't be a timetable, or even a series of milestones, but it should give a better picture of what we can expect as nanotechnology develops from simple, crude, costly beginnings to a state of greater sophistication and lower cost. Molecular manufacturing will make better products possible. We're likely to see some early applications in at least two areas: stronger materials and faster computers. Strong materials are simple, and will be hard to pass up. Computers are more complex, but the payoff will be enormous.

VARIOUS TYPES OF STRUCTURES

At the opposite extreme from molecular electronics-complex and at first worth billions of dollars per gram-are structural materials: worth only dollars per kilogram in most applications, but much simpler in structure. Once molecular manufacturing becomes inexpensive, structural materials will be important products.

These materials play a central role in almost everything around us, from cars and aircraft to furniture and houses. All of these objects get their size, shape, and strength from a structural skeleton of some sort. This makes structural materials a natural place to begin in understanding how nanotechnology can improve products.

Cars today are mostly made of steel, aircraft of aluminum, and buildings and furniture largely of steel and wood. These materials have a certain "strength-to-weight ratio" (more properly, a strength-to-density ratio). To make cars stronger, they'd have to be heavier; to make them lighter, they'd have to be weaker. Clever design can change this relationship a little, but to change it a lot requires a change of materials.

Making something heavy is easy: just leave a hollow space, then fill it with water, sand, or lead shot. Making something light and strong is harder, but often important. Automakers try to make cars lightweight, aircraft manufacturers try harder, and with spacecraft manufacturers it is an obsession. Reducing mass saves materials and energy.

The strongest materials in use today are mostly made of carbon. Kevlar, used in racing sails and bulletproof vests, is made of carbon-rich molecular fibers. Expensive graphite composites, used in tennis rackets and jet aircraft, are made using pure-carbon fibers. Perfect fibers of carbon-both graphite and diamond-would be even better, but can't be made with today's technology. Once molecular manufacturing gets rolling, though, such materials will be commonplace and inexpensive.

What will these materials be like? To picture them, a good place to start is wood. The structure of wood can vary from extremely light and porous, like balsa wood, to denser structures like oak. Wood is made by molecular machinery in plants from carbon-rich polymers, mostly cellulose. Molecular manufacturing will be able to make materials like these, but with a strength-to-weight ratio about a hundred times that of mediocre steel, and tens of times better than the best steel.

Instead of being made of cellulose, these materials will be made of carbon in forms like diamond. Diamond is emphasized here not because it is shiny and expensive, but because it is strong and potentially cheap. Diamond is just carbon with properly arranged atoms. Companies are already learning to make it from natural gas at low pressure. Molecular manufacturing will be able to make complex objects of the stuff, built lighter than balsa wood but stronger than steel.

Products made of such materials could be startling by our present standards. Objects could be made that are identical in size and shape to those we make today, but simultaneously stronger and 90 percent lighter. This is something to keep in mind next time you're lugging a heavy object around. (If something needs weight to hold it in place, it would be more convenient to add this ballast when the thing is in its proper location than to build in the extra weight permanently.) Better structural materials will make aircraft lighter, stronger, and more efficient, but will have the greatest effect on spacecraft. Today, spacecraft can barely reach orbit with both a safety margin and a cargo. To get there at all, they have to drop off parts like boosters and tanks along the way, shedding weight.

FAST DEVELOPMENT

In some areas of high technology-spaceflight has been a notorious example-it takes years, even decades, to try a new idea. This makes progress slow to a crawl. In other areas-software has been a shining example-new ideas can be tested in minutes or hours. Since the Space Shuttle design was frozen, personal computer software has come into existence and gone through several generations of commercial development, each with many cycles of building and testing.

Fast, Inexpensive Testing

Even in the days of the first operational molecular manipulators, experimentation is likely to be reasonably fast. Individual chemical steps can take seconds or less. Complex molecular objects could be built in a matter of hours. This will let new ideas be put into practice almost as fast as they can be designed. Later assemblers will be even faster. At a millionth of a second per step, they will approach the speed of computers. And, as nanotechnology matures, experimenters will have more and more molecular instruments available to help them find out whether their devices work or not. Fast construction and fast testing will encourage fast progress.

At this point, the cost of materials and equipment for experiments will be trivial. No one today can afford to build Moon rockets on a hobby budget, but they can afford to build software, and many useful programs have been the result. There is no economic reason why nanomachines couldn't eventually be built with a hobby-size budget, though there are reasons-to be discussed in later chapters-for wanting to place limits on what can be built.

Prior Simplicity

Finally, established technologies are always pushing up against some limit; the easy opportunities have generally been exploited. In many fields, the limits are those of the properties of the materials used and the cost and precision of manufacturing. This is true for computers, for spacecraft, for cars, blenders, and shoes.

For software, the limits are those of computer capacity and of sheer complexity (which is to say, of human intelligence). After molecular manufacturing develops certain basic abilities, a whole set of limits will fall, and a whole range of developments will become possible. Limits set by materials properties, and by the cost and precision of manufacturing, will be pushed way back. Competition, easy opportunities, and fast, low-cost experimentation should combine to yield an explosion of new products.

This does not mean immediately, and it does not apply to all imaginable nanotechnologies. Some technologies are imaginable and clearly feasible, yet dauntingly complex. Still, the above considerations suggest that a wide range of advances could happen at a brisk pace. The main bottleneck might seem to be a shortage of knowledgeable designers-hardly anyone knows both chemistry and mechanical design-but improving computer simulations will help. These simulations will let engineers tinker with molecular-machinery designs, absorbing knowledge of chemical rules without learning chemistry in the usual sense.

Climbing Complexity

Making familiar products from improved materials will increase their safety, performance, and usefulness. It will also present the simplest engineering task. A greater change, though, will result from unfamiliar products made possible by new manufacturing methods. In talking about unfamiliar products, a hard-to-answer question arises: What will people want?

Products are typically made because their recipients want them. In our discussions here, if we describe something that people won't want, then it probably won't get built, and if it does get built, it will soon disappear. (The exceptions-fraud, coercion, persistent mistakes-are important, but in other contexts.) To anchor our discussion, it makes sense to look not at totally new products, but instead at new features for old products, or new ways to provide old services.

This approach won't cover more than a fraction of what is possible, but will start from something sensible and provide a springboard for the imagination. As usual, we are describing possibilities, not making predictions. The possibilities focused on here arise from more complex applications of molecular manufacturing-nanotechnological products that contain nanomachines when they are finished. Earlier, we discussed strong materials. Now, we discuss some smart materials.

The goal of making materials and objects smart isn't new: researchers are already struggling to build structures that can sense internal and environmental conditions and adapt themselves appropriately. There is even a Journal of Intelligent Material Systems and Structures. By using materials that can adapt their shapes, sometimes hooked up to sensors and computers, engineers are starting to make objects they call "smart." These are the early ancestors of the smart materials that molecular manufacturing will make possible.

Today, we are used to having machines with a few visible moving parts. In cars, the wheels go around, the windshield wipers go back and forth, the antenna may go up and down, the seat belts, mirrors, and steering wheel may be motor-driven. Electric motors are fairly small, fairly inexpensive, and fairly reliable, so they are fairly common. The result is machines that are fairly smart and flexible, in a clumsy, expensive way.

In the Desert Rose scenario, we saw "tents" being assembled from trillions of submicroscopically small parts, including motors, computers, fibers, and struts. To the naked eye, materials made from these parts could seem as smooth and uniform as a piece of plastic, or as richly textured as wood or cloth-it is all a matter of the arrangement and appearance of the submicroscopic parts.

These motors and other parts cost less than a trillionth of a dollar apiece. They can be quite reliable, and good design can make systems work smoothly even if 10 percent of a trillion motors burn out. Likewise for motor-controlling computers and the rest. The resulting machines can be very smart and flexible, compared to those of today, and inexpensive, too. When materials can be full of motors and controllers, whole chunks of material can be made flexible and controllable. The applications should be broad.

Surfaces surround us, and human-made surfaces-walls, roofs, and pavement-cover huge areas that matter to people. How can smart materials make a difference here?

The revolution in technology has come and gone, and you want to repaint your walls. Breathing toxic solvents and polluting water by washing brushes have passed into history, because paint has been replaced with smarter stuff. The mid-twentieth century had seen considerable progress in paints, especially the development of liquids that weren't quite liquid-they would spread with a brush, but didn't (stupidly) run and drip under their own weight. This was an improvement, but the new material, "paperpaint," is even more cooperative.

Paperpaint comes in a box with a special trowel and pen. The paperpaint itself is a dry block that feels a lot like a block of wood. Following the instructions, you use the pen to draw a line around the edge of the area you want to paint, putting an X in the middle to show where you want the paint to go on; the line is made of nontoxic disappearing ink, so you can slop it around without staining anything. Using the trowel, you slice off a hunk of paperpaint-which is easy, because it parts like soft butter to the trowel, even though it behaves like a solid to everything else. Very high IQ stuff, that.

Next, you press the hunk against the X and start smoothing it out with the trowel. Each stroke spreads a wide swath of paperpaint, much wider than the trowel, but always staying within the inked line. A few swipes spreads it precisely to the edges, whereupon it smooths out into a uniform layer. Why doesn't it just spread itself? Experience showed that customers didn't mind the effort of making a few swipes and preferred the added control. The paperpaint consists of a huge number of nanomachines with little wheels for rolling over one another and little sticky pads for clinging to surfaces. Each has a simple, stupid computer on board. Each can signal its neigbours.

The whole mass of them clings together like an ordinary solid, but they can slip and slide in a controlled way when signaled. When you smooth the trowel over them, this contact tells them to get moving and spread out. When they hit the line, this tells them to stop. If they don't hit a line, they go a few handbreadths, then stop anyway until you trowel them again. When they encounter a line on all sides, word gets around, and they jostle around to form a smooth, uniform layer.

Any that get scraped off are just so much loose dust, but they stick together quite well. This paint-stuff doesn't get anything wet, doesn't stain, and clings to surfaces just tightly enough to keep it from peeling off accidentally. If some experimentally minded child starts digging with a stick, makes a tear, and peels some off, it can be smoothed back again and will rejoin as good as new. The child may eat a piece, but careful regulation and testing has ensured that this is no worse than eating plain paper, and safer than eating a colorful Sunday newspaper page.

Many refinements are possible. Swipes and pats of the trowel could make areas thicken or thin, or bridge small holes (no more Spackling!). With sufficiently smart paperpaint, and some way to indicate what it should do, you can have your choice of textures. Any good design will be washable, and a better design would shed dirt automatically using microscopic brushes. Removal, of course, is easy: either you rip and peel (no scraping needed), or find that trowel, set the dial on the handle to "strip," and poke the surface a few times. Either way, you end up with a lump ready to pitch into the recycling bin and the same old wall you started with, bared to sight again. Looking further at the human environment we find a lot of cloth and related materials, such as carpeting and shoes. The textile industry was at the cutting edge of the first industrial revolution, and the next industrial revolution will have its effects on textiles.

With nanotechnology, even the finest textile fibers could have sensors, computers, and motors in their core at little extra cost. Fabrics could include sensors able to detect light, heat, pressure, moisture, stress, and wear, networks of simple computers to integrate this data, and motors and other nanomechanisms to respond to it. Ordinary, everyday things like fabric and padding could be made responsive to a person's needs-changing shape, color, texture, fit, and so forth-with the weather and a person's posture or situation.

This process could be slow, or it could be fast enough to respond to a gesture. One result would be genuine one-size-fits-all clothing (give or take child sizes), perfectly tailored off the rack, warm in winter, cool and dry in summer; in short, nanotechnology could provide what advertisers have only promised. Even bogus advertising gives a clue to human desires.

Throughout history, the human race has pursued the quest for comfortable shoes. With fully adjustable materials, the seemingly impossible goal of having shoes that both look good and feel good should finally be achieved. Shoes could keep your feet dry, and warm except in the Arctic, cool except in the tropics, and as comfortable as they can be with a person stepping on them.

Adaptive structures will be useful in furniture. Today, we have furniture that adapts to the human body, but it does so in an awkward and incomplete manner. It adapts because people grab cushions and move them around. Or a chair adapts because it is a hinged contraption that grudgingly bends and extends in a few places to suit a small range of preferred positions. Occasionally, one sees furniture that allegedly gives a massage, but in fact only vibrates.

These limitations are consequences of the expense, bulkiness, clumsiness, and unreliability of such things as moving parts, motors, sensors, and computers today. With molecular manufacturing, it will be easy to make furniture from smart materials that can adapt to an individual human body, and to a person's changing position, to consistently give comfortable support. Smart cushions could also do a better job of responding to hints in the form of pats, tugs, and punches. As for massage-a piece of furniture, no matter how advanced, is not the same as a masseuse. Still, a typical massage setting on a smart chair would not mean today's "vibrate medium vigorously," but something closer to "five minutes of shiatsu."

This tour through of the potential of smart matter has shown how we could get walls that look and sound as we wish, clothing, shoes, and furniture of greater comfort, and clean solar power. As one might expect, this just scratches the surface.

If you care to think of further applications, here are some ground rules: Components made by molecular manufacturing can be many tens of times stronger than steel, but materials made by plugging many components together will be weaker. For these, strengths in the range of cotton candy to steel seem achievable. The components will be sensitive to heat, and at high temperatures they will break down or burn. Many materials will be able to survive the temperature of boiling water, but only specialized designs would be oven-safe.

Color, texture, and (usually) sound should be controllable. Surfaces can be smooth and tightly sealed (this takes some cleverness). Motions can be fairly fast Power has to come from somewhere; good sources include electricity, stored chemical energy, and light. If nanomachines or smart materials are dunked in liquids, chemical energy can come from dissolved molecules; if they are in the open, energy can come from light; if they are sitting in one place, they can be plugged into a socket; if they are moving around in the dark, they can run on batteries for a while, then run down and quit. Within these limits, much can be accomplished.

"Smart" is a relative term. Unless you want to assume that people learn a lot more about intelligence and programming, it is best to assume that these materials will follow simple rules, like those followed by parts of drawings on computer screens. In these drawings, a picture of a rectangle can be commanded to sprout handles at its corners; pulling a handle stretches or shrinks the rectangle without distorting its right-angle corners. An object made of smart matter could do likewise in the real world: a box could be stretched to a different size, then made rigid again; a door in a smart-material wall could have its position unlocked, its frame moved a pace to the left, and then be returned to normal use.

There seems little reason to make bits of smart matter independent, self-replicating, or toxic. With care, smart matter should be safer than what it replaces because it will be better controlled. Spray paint gets all over things and contains noxious solvents; the paperpaint described above doesn't. This will be a characteristic difference, if we exercise our usual vigilance to encourage the production of things that are safe and environmentally sound.

It may be fun to discuss wondrous new products, but they won't make much difference in the world if they are too expensive. Besides, many people today don't have decent food, clothes, and a roof over their heads, to say nothing of fancy "nanostuff."

Costs matter. There is more to life than material goods, but without material goods life is miserable and narrow. If goods are expensive, people strive for them; if goods are abundant, people can turn their attention elsewhere. Some of us like to think that we are above a concern for material goods, but this seems more common in the wealthy countries. Lowering manufacturing costs is a mundane concern, but so are feeding people, housing them, and building sewage systems to keep them from dying of cholera and hepatitis. For all these reasons, finding ways to bring down production costs is a worthy goal. For the poor, for the environment, and for the freeing of human potential, costs matter deeply. Let's take a closer look at the costs of molecular manufacturing.

Inflation produces the illusion that costs rise, when the real story is that the value of money is falling. In the short term, real costs usually don't change very quickly, and this can produce the illusion that costs are stable facts of nature, like the law of gravity or the laws of thermodynamics. In the real world, though, most costs have been falling by a crucial measure: the amount of human labor needed to make things. People can afford more and more, because their labor, supplemented by machines, can produce more and more. This change is dramatic measured on a scale of centuries, and equally dramatic across the gulf between Third World and developed countries. The rise from Third World to First World standards of living has raised income (dropped the cost of labor time) by more than a factor of ten. What can molecular manufacturing do?

Larger cost reductions have happened, most dramatically in computers. The cost of a computer of a given ability has fallen by roughly a factor of 10 every seven years since the 1940s. In total, this is a factor of a million. If automotive technologies had done likewise, a luxury car would now cost less than one cent. (Personal computer systems still cost hundreds of dollars both because they are far more powerful than the giant machines of the 1940s and because the cost of buying any useful computer system includes much more than just the cost of a bare computer chip.)

Bibliography

Attebery, B.: *Nanoculture: Implications of the New Technoscience*, Bristol, Intellect Books, 2004.

Bainbridge, W.S. : *Societal Implications of Nanoscience and Nanotechnology*, Kluwer, Dordrecht, 2001.

Berube, D.M.: *The Rhetoric of Nanotechnology*, Amsterdam: IOS Press, 2004.

Charles, P., and Frank J. Owens: *Introduction to Nanotechnology*, Hoboken, J. Wiley, 2003.

Credi, A.: *Molecular Devices and Machines: a Journey into the Nano World*, Weinheim, Wiley-VCH, 2003.

Daniel, Ratner: *Nanotechnology: a Gentle Introduction to the Next Big Idea*, Upper Saddle River, Prentice Hall, 2003.

Drexler, K. Eric: *Engines of Creation: The Coming Era of Nanotechnology*, New York, Anchor Books, 1986.

Drexler, K. Eric: *Unbounding the Future: the Nanotechnology Revolution*, New York, Morrow, 1991.

Frank J. Owens: *Introduction to Nanotechnology*, Hoboken, J. Wiley, 2003.

Glimell, H.: *Dynamics of the Emerging Field of Nanoscience*, Dordrecht, Kluwer, 2001.

Gregory, L.: *Nanotechnology*, New York, Springer, 1999.

Hayles, N.K. : *Nanoculture: Implications of the New Technoscience*, Bristol, Intellect Books, 2004.

Hessenbruch, A.: *Nanotechnology and the Negotiation of Novelty*, Amsterdam, IOS Press, 2004.

Mee, W: *Citizens' Panel on Nanotechnology: Report to Participants,* Melbourne, CSIRO Minerals, 2005.

Pressman, J.: *Nano Narrative: A Parable from Electronic Literature*, Bristol, Intellect Books, 2004.

Rietman, Ed.: *Molecular Engineering of Nanosystems,* New York, Springer, 2001.

Sarewitz, D.: *Nanotechnology and Societal Transformation*, Dordrecht, Kluwer, 2001.

Schummer, J.: *Interdisciplinary Issues in Nanoscale Research*, Amsterdam: IOS Press, 2004.

Seeman, N.: *DNA Nanotechnology for Massive Information Storage,* New York, New York University, 2001.

Shew, A.: *Discovering the Nanoscale*, Amsterdam: IOS Press, 2004.

Smith, H. A.: *Nanotechnology Infrared Optics for Astronomy Missions,* Washington, National Aeronautics and Space Administration, 2002.

Stephen M. Goodnick: *Transport in Nanostructures,* Cambridge, Cambridge University Press, 1997.

Stephen M. Goodnick: *Transport in Nanostructures,* Cambridge, Cambridge University Press, 1997.

Suchman, M.: *Envisioning Life on the Nano-Frontier*, Dordrecht, Kluwer, 2001.

Tenner, E.: *Nanotechnology and Unintended Consequences*, , Dordrecht, Kluwer, 2001.

Tomellini, R.: *Nanotechnology: Revolutionary Opportunities and Societal Implications, Workshop, Lecce*, Luxemburg: Office for Official Publications of the European Communities, 2002.

Index

F

I

L

M

O

P

R

S

T

U

V